The Paradox

'More than a tourist destination, Svalbard is a hotspot of geopolitics, climate change, transient migration and social inequalities. Engaging, rich and nuanced, this book gives voice to people whose stories are rarely told, and exposes the deep dilemmas facing this Arctic archipelago. This book is a must for anyone with an interest in Svalbard, and the challenges of a melting world. Ethnography at its best.'

—Marianne E. Lien, Professor, Department of Social Anthropology,
University of Oslo

'In a rich and deeply-textured account of the human communities that call Svalbard "home", Zdenka Sokolíčková demonstrates how the logic of extraction intersects awkwardly with community, environment, geopolitics and sustainability. If Svalbard is a paradox then it will demand explicit recognition of the competing interests, pressures and wishes that make the archipelago and its communities such intriguing places to live, work and study.'

—Klaus Dodds, Professor of Geopolitics, Royal Holloway
University of London

'Lucidly captures the dilemmas of maintaining community in the world's northernmost settlement, where climate change is particularly evident. Through fine-grained ethnography, this weaves together questions of belonging, labour and inequality with the paradoxes of "green growth" initiatives and geopolitics. Highly recommended!'

—Cecilie Vindal Ødegaard, Professor of Social Anthropology,
University of Bergen

'Sokolíčková profoundly and poetically reveals Svalbard as a site of concentrated uncertainty: simultaneously microcosm and periphery, container for a range of peculiarly twenty-first-century meanings, and home to a community unique in the world.'

—Adam Grydehøj, editor-in-chief of *Island Studies* journal

Anthropology, Culture and Society

Series Editors:
Holly High, Deakin University
and
Joshua O. Reno, Binghamton University

Recent titles:

The Paradox of Svalbard

Climate Change and Globalisation in the Arctic

Zdenka Sokolíčková

Foreword by Thomas Hylland Eriksen

First published 2023 by Pluto Press
New Wing, Somerset House, Strand, London WC2R 1LA
and Pluto Press, Inc.
1930 Village Center Circle, 3-834, Las Vegas, NV 89134

www.plutobooks.com

British Library Cataloguing in Publication Data
A catalogue record for this book is available from the British Library

ISBN 978 0 7453 4740 0 Paperback
ISBN 978 0 7453 4743 1 PDF
ISBN 978 0 7453 4742 4 EPUB

This book is printed on paper suitable for recycling and made from fully
managed and sustained forest sources. Logging, pulping and manufacturing
processes are expected to conform to the environmental standards of the
country of origin.

Typeset by Stanford DTP Services, Northampton, England

Simultaneously printed in the United Kingdom and United States of America

Contents

Figures

Abbreviations

EU	European Union
FiFos	fly in-fly out workers
FTE	full-time equivalent (measure of available jobs)
LL	Longyearbyen Lokalstyre (Svalbard version of a municipal council)
LO	Norwegian Confederation of Trade Unions
NGO	non-governmental organisation
NOK	Norwegian kroner (currency)
R&D	research and development
SNSK	Store Norske Spitsbergen Kulkompani (the Norwegian Coalmining Company of Spitsbergen)
UNIS	University Centre in Svalbard

Series Preface

As people around the world confront the inequality and injustice of new forms of oppression, as well as the impacts of human life on planetary ecosystems, this book series asks what anthropology can contribute to the crises and challenges of the twenty-first century. Our goal is to establish a distinctive anthropological contribution to debates and discussions that are often dominated by politics and economics. What is sorely lacking, and what anthropological methods can provide, is an appreciation of the human condition.

We publish works that draw inspiration from traditions of ethnographic research and anthropological analysis to address power and social change while keeping the struggles and stories of human beings centre stage. We welcome books that set out to make anthropology matter, bringing classic anthropological concerns with exchange, difference, belief, kinship and the material world into engagement with contemporary environmental change, capitalist economy and forms of inequality. We publish work from all traditions of anthropology, combining theoretical debate with empirical evidence to demonstrate the unique contribution anthropology can make to understanding the contemporary world.

Holly High and Joshua O. Reno

Acknowledgements

The path that led to this book started in 2016 with the idea of exploring whether Longyearbyen in Svalbard was a suitable place to learn something new about overheating. When Thomas Hylland Eriksen from the Department of Social Anthropology at the University of Oslo in Norway encouraged me to apply for funding, the first milestone was achieved: I decided to give it a try. It was one year before that, in 2015, when Thomas and I were sitting together at Tromsø Airport, that I asked him what it feels like to be an anthropological legend. 'You know what, I am just an ordinary guy,' Thomas replied. He is not. Had he not believed an unknown Czech female scholar, with three small kids and an accordingly poor publication record could do meaningful ethnography in the peculiar town at 78° North, *The Paradox of Svalbard* would not have come into existence.

The second entity that deserves to be given merit is an anonymous assemblage – the fluid, multiscalar and superdiverse community of Longyearbyen consisting of innumerable generous individuals who saw a value in contributing to my study, be it as private person or as representatives of various institutions. It is not possible to name them all as there were literally a couple of hundred of them. I am grateful for all the moments when they shared their life stories, reflected on the questions I posed and nudged me to look at things from different angles than I anticipated.

From the large pool of my interlocutors, several friendships developed and made my stay in Longyearbyen joyful in a way that stretches far beyond the academic achievement of publishing an ethnographic monograph. The warmheartedness of Hilde Henningsen, Ingvild Sæbu Vatn, Lilli Wickström, Lilith Kuckero, Mark Sabbatini, Dagmara Wojtanowicz, but also the women's choir *Tundradundrene*, my children's teachers and parents of their friends who welcomed us and embraced us as part of Longyearbyen's transient mesh will not be forgotten.

My thinking about what is happening in Longyearbyen would have been much poorer had I not had the opportunity to exchange ideas with my colleagues, social scientists, artists and humanities scholars who gather under the umbrella of the Svalbard Social Science Initiative. Discussions thanks to which I could sharpen my argument – both through verifying that others analyse the processes under way in Longyearbyen along similar lines and through contemplating competing views of events played an important role during my fieldwork – continued after I left Svalbard and will hopefully unfold further after *The Paradox of Svalbard* is published.

The very manuscript of the book was reviewed by several people I would like to acknowledge for the selfless feedback they provided, giving me concrete and constructive suggestions as to how to boost the message my monograph wishes to convey. A special mention here goes to Gabrielle Hecht, Alexandra Meyer, Tomas Salem, Trine Andersen and the publisher's anonymous reviewers. I would also like to thank all my colleagues who offered feedback on earlier drafts of the journal articles and book chapters I have so far published about Longyearbyen, and their peer reviewers; in the writing process, ideas crystallise over time and there is no way how to speed it up. Thank you for being patient with me and believing I can improve.

When my fieldwork ended in 2021, I got a new job at the Arctic Centre at the University of Groningen in the Netherlands. I am grateful to René van der Wal and Maarten Loonen for giving me space to finish the work on my book. Thanks also to my Arctic Centre colleague Esteban Ramirez Hincapié for providing me with the software of Scrivener, which made the technical aspect of writing and revising such a substantial chunk of text as a monograph incredibly smooth.

The whole team of Pluto Press deserves to be acknowledged for their diligence and accuracy. A special mention goes to my editor David Castle, who believed in the book since I sent my first email to him in 2020.

The University of Hradec Králové administered and co-funded the research project CZ.02.2.69/0.0/0.0/18_070/00094 76 (bo)REALIFE: Overheating in the High Arctic: Qualitative Anthropological Analysis, financed by the European Union through the Ministry of Education, Youth and Sports of the Czech Republic. I appreciate that both my home university and the funder trusted in my capability to carry out the project and supported me during several years of fieldwork, as well as during the many months of writing. I also value highly the encouragement of Tomáš Petráček, the Head of the Department of Studies in Culture and Religion where I worked as a teaching associate before I moved to Svalbard; Tomáš knew he was losing a team member for a significant period of time but he supported me in my decision to do what I believed made sense.

When I was finishing my work on the manuscript, there were moments of despair and exhaustion. I would like to thank Line Nagell Ylvisåker for her reassuring messages that I could make it. 'Breathe,' she said, 'and smell the kids.' I did, and it helped. Immense thanks to my friends, my husband's family and my husband Jakub Žárský for supporting me in the idea to move to Longyearbyen for a couple of years. Thanks also to our sons Josef, Vratislav and Adam for not objecting. Jakub and the boys had to stretch quite a bit in their patience and efforts to cheer me up while I was struggling with the pains of producing something as monumental as a book. Nothing can be achieved unless others who care about you make way for your dreams to be realised.

*   *   *

Parts of the book were published in the form of journal articles and book chapters, as listed below. I also indicate in brackets which chapter in the book includes the reused passages. I am grateful to the publishers for allowing me to reproduce the texts within this book.

- Sokolíčková, Z. (2022). The golden opportunity? Migration to Svalbard from Thailand and the Philippines. *Nordic Journal of Migration Research* 12(3): 293–309. (Chapter 8)
- Sokolíčková, Z. (2022). The trouble with local community in Long-yearbyen, Svalbard: How *big politics* and lack of *fellesskap* hinder a not-yet-decided future. *Polar Record* 58, E36. (Chapter 7)
- Sokolíčková, Z. and Eriksen, T.H. (2023). Extraction cultures in Svalbard: From mining coal to mining knowledge and memories. In Sörlin, S. (Ed.). *Resource Extraction and Arctic Communities: The New Extractivist Paradigm*, pp. 66–86. Cambridge: Cambridge University Press. (Introduction; Chapter 4)

Further publications where I, alone or with colleagues, draw on the results of my fieldwork and where an interested reader can gain insight into topics hinted at in the book include:

- Sokolíčková, Z. and Soukupová, E. (2021). Czechs and Slovaks in Svalbard: Entangled modes of mobility, place and identity. *Urban People* 23(2): 167–196.
- Sokolíčková, Z. (2022). The Chinese riddle: Tourism, China and Svalbard. In Lee, Y.-S. (ed.) *Asian Mobilities Consumption in a Changing Arctic*, pp. 141–154. Abingdon: Routledge.

- Sokolíčková, Z., Meyer, A. and Vlakhov, A. (2022). Changing Svalbard: Tracing interrelated socioeconomic and environmental change in remote Arctic settlements. *Polar Record* 58, E23.
- Brode-Roger, D., Zhang, J., Meyer, A. and Sokolíčková, Z. (2022). Caught in between and in transit: Forced and encouraged (im)mobilities during the Covid-19 pandemic in Longyearbyen, Svalbard. *Geografiska Annaler: Series B, Human Geography.* https://doi.org/10.1080/04353684.2022.2097937
- Sokolíčková, Z., Ramirez Hincapié, E., Zhang, J., Lennert, A.E., Löf, A. and van der Wal, R. (2023). Waters that matter: How human–environment relations are changing in high-Arctic Svalbard. *Anthropological Notebooks* 28(3): 74–109.
- Meyer, A. and Sokolíčková, Z. (forthcoming). Melting worlds and climate myths: Stories of climate change in Longyearbyen, an Arctic 'frontline community'. *Ethnos.*

Publicly available are my conference talks presented remotely, where I sketch my early ideas on the issues discussed in the book:

- The northernmost dilemma: Perception of climate change in the post-mining 'melttown' of Longyearbyen. EASA2020, Lisbon, 22 July 2020, https://www.youtube.com/watch?v=s3zbWMP2TL0
- 'My world is melting': Perception of environmental change in Longyearbyen, Svalbard. Together with Thomas H. Eriksen and Line N. Ylvisåker, VANDA Anthropology Days 2020, Vienna, 29 September 2020, https://www.youtube.com/watch?v=qSDUlc6NMPk
- The ethics of sharing in Longyearbyen. At 'The Ethics of Sharing', Aarhus, 11 May 2021, https://www.youtube.com/watch?v=-BMnJU T2Yc8

Foreword

Thomas Hylland Eriksen

There are many ways of approaching Zdenka Sokolíčková's rich and engaging monograph, and I shall start by taking the easy way out, namely by highlighting one perspective from her book. It is perhaps the most obvious one, but it is also one that contributes to the scaffolding and framing of the entire project. It is typically in small communities embedded in global networks that the central contradictions of contemporary humanity are at their most visible. In Longyearbyen (pop. 2,400), global inequality, racist exclusion, climate change and the impossible challenge of a green transition without systemic change are transparent and impossible to ignore.

This is not to say that Longyearbyen is a microcosm of anything else. Scale matters, and scaling up or down leads to structural transformations rather than more or less of the same thing. An idea occasionally promoted in the tourist industry, dismissed by Sokolíčková, to the effect that Svalbard somehow is a miniature of Norway, which in turn is a miniature of the world, is patently absurd. Norway is cold, rich and thinly populated. Its oil wealth has insulated it from the most severe effects of global crises (tellingly, *austerity* has no precise equivalent in Norwegian). Norwegians could watch, from a distance, the catastrophic floodwaters from China to California and the devastating European droughts, no longer confined to the Mediterranean, during the uncannily unusual summer of 2022. Although consumer prices have increased by percentages in the double digits across the continent, including Norway, Norwegians are unlikely to starve in the foreseeable future. Although they rely on high energy consumption for reasons of lifestyle and climate, they can afford their electricity and petrol. Indeed, they are themselves major producers of both, thanks to the abundant availability of hydroelectric power and North Sea oil, which transformed Norwegian society in the latter decades of the last century.

Shifting the gaze a notch down to Svalbard, it is hardly a 'Norway in miniature' although it can perhaps be seen as a hardcore, extreme version of the mountain massifs of the mainland. Notably, there was no Indigenous population in the frozen (but slowly thawing) archipelago, which is incapable of supporting a human population without external input. So Svalbard simultaneously epitomises the double binds and inequalities of twenty-first century modernity *and* is in several ways unique.

Several features of Svalbard makes it unusual, indeed globally unique. As Sokolíčková shows – and this is a major theme in the book – human activities in the archipelago have always been purely extractive. Visitors who courageously braved the extremes of the Arctic climate before the advent of Gore-Tex and central heating were mostly engaged in trapping, hunting or whaling. They took something out without giving anything back. From the beginning of the last century, coal mining took pride of place in the economy of Spitsbergen, thereby expanding the range of extractive activities.

The very presence of coal in the barren wastelands of the far north merits some reflection. The philosopher Jean-Paul Sartre once mused that coal, which gave us industrial society and modernity as we know it, could be seen as a gift bequeathed to humanity by organisms that lived and died aeons ago (Sartre 1960). A poetic and beautiful thought, it is inadequate at a time when fossil fuels are recognised as a double-edged, limited good but, writing in 1960, Sartre was not to blame. Another, no less mind-boggling perspective concerns the location of the coal. Most of the high-quality coal mined in Svalbard dates back to the appropriately named Carboniferous era, more than 300 million years ago, or the subsequent Permian. Svalbard was covered in dense, lush, tropical vegetation at the time; not because the world was much hotter than now, but because of continental drift. The islands now pointing towards the North Pole formed parts of continents in a different part of the planet. So when the official tourist board *Visit Svalbard*, a collaborative effort involving many smaller and slightly larger enterprises, invites tourists to visit Mine 3 in order to 'dig deep in the history of Longyearbyen, the miners, all the coal dust and their life in the mountains', their conceptualisation of deep history differs from that of the geologists, or of imaginative philosophers such as Sartre.

Change, in other words, is endemic to the planetary condition. What is disconcerting and frightening in the present century is not change as such, but the speed of global change coupled with a lack of control. The global condition can well be described as an overheated one, both literally and metaphorically: acceleration is accelerating in a number of domains, from world trade and amplification of lifeworlds to pollution and climate change, and there is no thermostat (or governor, in cybernetic parlance) enabling stability and long-term sustainability (Eriksen

2016). As everybody knows by now, literal overheating is the second defining characteristic of Svalbard: climate change, which makes itself felt everywhere in the world, has more dramatic immediate consequences in the Arctic and Antarctic than elsewhere, certainly if we choose to focus on global warming (although climate change is also expressed in other ways). The climate, we are repeatedly being told through news channels, is warming faster in the Arctic than anywhere else. The seal hunts of Inuit in Greenland and Canada are disrupted by the withdrawal and thinning of sea ice (Hastrup 2021); fish species migrate owing to warming seas, pulling the cod fisheries of Northern Norway further and further north. Unprecedented winter thaws necessitate manual feeding of reindeer among Sámi pastoralists, leading to a foraging crisis (*beitekrise*) and generating dark jokes among Sámi about starting to 'smell like farmers' because they handle bales of hay, day in day out, instead of warming themselves by an outdoor fire while their reindeer peacefully munch lichen nearby (Lien 2022). In Svalbard, the deadly avalanche described by Sokolíčková is a forewarning, as are the occasional January rains and record temperatures in July.

A third particularity of Svalbard, which was not foregrounded in Sokolíčková's original research project on the local implications of climate change, concerns its special status in international law and its implications for people coming to Svalbard from places other than Norway. Quite contrary to the tourist imagery of that pristine, white, rough place, Nordic in the extreme, some of the most numerous and most settled people in Longyearbyen have travelled from the Philippines and Thailand. It speaks volumes about global inequality that they eke out a living in an environment which is hostile in more than one way, but the most striking expression of hierarchical racialisation is the complete absence of Asians from tourist materials and stereotypically Arctic work functions such as tour guide service. To the foreign tourist gaze, they are as foreign as an African don at an Oxford college in the nineteenth century. Racialisation is not a shameful memory, but part of the living present.

A fourth feature of Svalbard which makes it a special place is the universal obligation to fend for yourself. It is technically impossible to be unemployed in Longyearbyen, which is an outlandish notion to Norwegians, who are accustomed to living in a welfare state which looks after them from cradle to grave. And to follow up on the latter, they say that you cannot really die in Longyearbyen either, since there is just one cemetery reserved for people with documented strong connections to the islands.

Obviously people do die at the most inconvenient times. So you have to be alive and to work in order to defend your place in the community. People live in Longyearbyen and, at the same time, they don't. As Sokolíčková points out, residents have not been equipped with ID cards documenting their place of residence.

Fifth and finally, its past status as *terra nullius*, an empty land to be conquered and exploited by colonisers, makes Svalbard unusual and different. Unlike in other parts of the Arctic, there were no native peoples to be subjugated (Asians may now be seen to play that part, belatedly). There were just wild, frozen expanses of wilderness, with scant vegetation even at the height of summer, a limited and fragile ecosphere with the polar bear at the apex, but also including other species such as Arctic foxes, ptarmigans and stunted reindeer in addition to diverse marine life, including many species of sea mammals. In our time, however, the ruthless exploitation of nature may come across as no less callous and insensitive than the colonisation of 'natives'. Since coal mining was never profitable in the first place, phasing it out has been relatively undramatic, although perceptibly tinged with social friction and cultural nostalgia.

An obvious follow-up question is why there should be a human population in Svalbard at all, since it is neither self-reliant, nor ecologically sustainable, nor a profit-making machine for the mainland. The short answer is a geopolitical one, but Sokolíčková's interlocutors have much more to say about the matter and, significantly, they do not speak with one voice. The shift to tourism and research, hailed by Norwegian authorities as a feasible alternative to mining, continues the extractive practice, as Sokolíčková points out.

*   *   *

This book highlights that which is unique to Svalbard, but its main analytical thrust lies in the way the author connects local concerns to global issues and the challenges experienced by comparable communities.

Regarded as a boomtown, Longyearbyen has much in common with similar communities elsewhere (Eriksen 2018). These locations typically grow, thrive and flourish for a few years or decades before rapidly going into decline. Boomtowns are communities with a finger on the fast-forward button. Typically built around a natural resource, they produce considerable prosperity until the resource is depleted. Some readers will inevitably think about the Klondike gold rush of the 1890s, but there are lots of twentieth-century and contemporary boomtowns which have gone through similar trajectories of boom and bust. They tend to be demographically dominated by working-class males – families are often fragile or settled elsewhere – and characterised by rapid turnover. As Sokolíčková shows, there are several men known as old-timers in Longyearbyen, but they would typically have moved back and forth to the mainland, and few, if any, have regarded the town as home all their life. The gender imbalance, paucity of stable nuclear families and rapid turnover rates indicate that building community in this sort of place requires hard continuous work. Perhaps the shift towards tourism and research, and the growing public administration of the town, will mitigate some of the boomtown effects, but scarcely all. The presence of schools and kindergartens shows that families are far from rare, and one of the latest initiatives from the mainland has consisted in the establishment of a junior college (*folkehøyskole*) in order to consolidate the Norwegian presence in a place which is simultaneously part of Norway and international territory.

Abandoned Australian boomtowns encountered during fieldwork down under have either been turned into outdoor museums, successfully or not, or been left to their own devices, which means dilapidation and decay. Before the modern obsession with the past, whaling stations and trapper's huts in Svalbard were also unceremoniously abandoned, akin to the Russian ghost town Pyramiden a short ride by snowmobile from Longyearbyen, which was left in great haste after the demise of the Soviet Union. As far as Longyearbyen is concerned, the authorities are determined to keep the flame burning, almost at any cost. Residents, be they old-timers or not, who have an emotional connection to the place, have genuinely mixed feelings. In one of the strongest sections in this book, residents with different agendas and interests express their views about cruise tourism, the expansion of UNIS (University Centre in Svalbard), climate change and the demise of mining. The contradiction between any desire to come across as green and ecologically responsible, and concurrent plans for expanding tourism, are glaring. Not just the actual tourists and the employees in the industry, but everything they consume has to be brought to the archipelago by plane or ship. The more people who visit or live in Svalbard, the greater the ecological footprint.

The contradiction between growth and sustainability, perhaps the most wicked of all problems facing humanity, is clearly evident in this transparent, fraught, changing, vulnerable society. At the same time, Sokolíčková's research confirms a view shared by many who have looked into these matters in other parts of the world: in general, climate change does not loom large in people's everyday life, even in a volatile Arctic town with an uncertain future. They have their jobs, their love relationships, their families, their holiday plans and their mortgages to worry about. They do, nevertheless, begin to worry when climate change impacts on their lives in disorienting and impractical ways. Perhaps one day, when the thaw and erratic weather patterns make tourism unattractive and Longyearbyen unliveable as permafrost segues into mud, the human consequences of climate change will inspire systemic change.

Yet there should be no place for wishful thinking in a scientific analysis of local responses to climate change. Steeped in an ideology and practice committed to the hegemonic ideology of growth and progress, Longyearbyen's residents are no different from most people living in capitalist societies by giving priority to the here and now rather than the there and then. The temporal scales in which people live are limited by their own lifespans and those of their children, their concerns of an urgent nature, their social lives embedded in the networks with which they are familiar. Yet, arguing for 'a long now' and 'a large here', transcending immediate concerns, is the bread and butter of ambitious ideologies and religions, some of which have met with considerable success in the past. Habituating oneself to thinking and acting beyond that which is close and immediate is far from impossible. Had it been, neither nationalism nor Christianity would have been possible, and yet both thrive in the very heart of modernity.

Most of Sokolíčková's interlocutors are conversant with the science of climate change and are perfectly aware that their lives are not feasible in the long term. A question often raised by researchers, not least those working in affluent societies with high literacy rates and universal education (e.g. Norgaard 2011), is, accordingly, why it is that people continue to live unsustainably while aware of the fact that they do so. One family of responses to this conundrum takes the concept of path dependence as its point of departure, arguing that people generally prefer to stick to their habits. A different family of explanations, which can be traced back not to behavioural economics but to classic thought – Plato comes to mind – assume the weakness of the will as a human characteristic. As St Paul phrased it succinctly: 'For I do not do what I want, but I do the very thing I hate' (Rom 7:15). In her extraordinarily rich and detailed life-world ethnography, Sokolíčková seems to suggest that unlike what many seem to believe, the problem is not a lack of knowledge or a poor 'public understanding of science'. In an important sense, the inadequate *scientific understanding of the public* is a greater problem. This is where Sokolíč-ková's ethnography truly shines, when she shows, convincingly and in rich detail, how people in Longyearbyen – like all of us – have to juggle contradictory expectations and conflicting values without becoming paralysed by cognitive dissonance. The figure of the miner, seen as a symbol of Svalbard, is a typical example. For decades the pride of the community, a symbol of hardship, wealth creation, masculinity and the collective efforts of the welfare state, the miner has recently suffered the same fate as the whaler, who can still be observed, in brass, on pedestals, in some coastal Norwegian cities, but who is now met with mixed feelings, even in the communities where whaling was a major source of prosperity for decades.

There is no easy way out of this double bind, no magical sword which can cut through the Gordian knot. Outsiders who learn about Norway and Svalbard may well ask why they, who live in societies where squalor and scarcity are the order of the day, should slow down and cool down for the sake of the climate, when an absurdly rich country almost devoid of social conflicts continues to expand its oil and gas industry. When President Bolsonaro defends logging in the Amazon, or President Tshisekedi decides to explore for oil in the Congolese rainforest, any objections from countries like Norway can easily be shrugged off. We are, after all, speaking of a country which cheerfully plans to open new oil fields in locations with vulnerable and fragile marine ecosystems, even if its citizens would have had food on their table anyway.

One of the great achievements of this monograph consists in its holistic approach to issues relating to global sustainability and climate change. Sokolíčková weaves this central issue into the broader social fabric of inclusion and exclusion, hierarchies and ethnic differences. Although the archipelago is frozen and barren, pitch dark for many weeks and generally dark for several months every year, her book is teeming with life. We meet a broad range of people, from miners to poets, from psychotherapists to cleaners to tourist guides, and the book invites the reader to join its author in a journey through the multiple facets of this surprisingly complex community. Just remember to bring lots of money and warm clothes.

*   *   *

There may be no other option than 'staying with the trouble' (Haraway 2016), but as Sokolíčková and several of her interlocutors indicate, there

xx   The Paradox of Svalbard

are alternatives, and abandonment is not the only one, even if it might have been the most sustainable one. Her interpretation of the *ouroboros* – the mythical snake eating its own tail – splendidly condenses the lead argument. Longyearbyen is not just a materially affluent but existentially struggling community at the edge of the world, it is also a human community in a world teetering on the edge of catastrophe. Sokolíčková has shared the stories of this community, and for this we should be grateful. What future generations will make of this book is impossible to predict, but my hunch is that mid-century readers are likely to see it as a prescient snapshot from a bygone age, an age of transition, a period when humanity lived briefly through the best of times and the worst of times.

References

Eriksen, T.H. (2016). *Overheating: An Anthropology of Accelerated Change*. London: Pluto.

—— (2018) *Boomtown: Runaway Globalisation on the Queensland Coast*. London: Pluto.

Haraway, D. (2016) *Staying with the Trouble: Making Kin in the Chthulucene*. Durham, NC: Duke University Press.

Hastrup, K. (2021) The end of nature? Inughuit life on the edge of time. *Ethnos* 88(1): 13–29, doi: 10.1080/00141844.2020.1853583

Lien, M.E. (2022) Climate change or a crisis of the commons? Sámi land use in the Anthropocene. Paper presented at the workshop 'Future Commons of the Anthropocene', EASA (European Association of Social Anthropologists) biannual conference, Belfast, 28 July.

Norgaard, K.M. (2011) *Living in Denial: Climate Change, Emotions, and Everyday Life*. Cambridge, MA: MIT Press.

Sartre, J.-P. (1960) *Critique de la raison dialectique*. Paris: Gallimard.

Introducing the Fieldwalk:
Field, Companions and Path

Svalbard as a Miniature of the World?

'Svalbard is a miniature of Norway. And Norway is a miniature of the world.' In November 2021, Jenny Skagestad, a Norwegian politician specialising in environment and transportation, and adviser in the environmental foundation ZERO, published a TEDx Talks video entitled 'Svalbard – Canary in the coal mine goes green'. In the annotation, we read:

> Svalbard is a climate paradox: A community next to the North Pole already threatened by the dramatic climate crisis and totally dependent on coal and diesel. But what happens when people start to ask new questions and challenge old systems? How can Svalbard become the showcase for both the climate crisis and the inspiring climate solutions?

Anthropology bewares of generalisations. Can we claim Svalbard teaches us anything about the world we inhabit? The meme of 'what happens in the Arctic does not stay in the Arctic' has been around for about a decade now, and it is usually associated with either climate change, or geopolitics and security. But can the statement also kick off a discussion about other issues relevant in the Arctic, such as globalisation, migration or social justice? How serious are we about asking fundamental questions? What does it entail to challenge old systems? And can a place be both a showcase of a crisis and its solutions? What does the paradoxical nature of multi-layered processes under way in Svalbard look like from within?

Longyearbyen, the biggest settlement on Svalbard, is a contradiction indeed. A living dilemma of the twenty-first century. An anthropological Petri dish where both climate change and globalisation 'are happening', they are fast and can be experienced by all five senses. You can see the glaciers diminishing and the faces of strangers – soon to become 'locals' – appearing. You can hear heavy machines working, numerous languages being spoken, gigantic cruise ships sounding their horns in the harbour, geese arriving and planes landing and taking off. You can feel raindrops on your face in December and polar bear fur with your fingers in front of a popular tourist store, inviting you to engage with a 'Touch me' sign. You can taste Thai spring rolls produced by a 'local', beef imported from New Zealand, crystal clean iceberg water 750 ml for €80 and coal particles carried by the blizzard squeaking between your teeth. And the smell; the smell is a tricky one. Svalbard first seems to have no smell. But there are scents. A Svalbardianer with a trained nose might be able to smell snow and ice. The kennels smell, especially when mild weather comes and thaws what is to be thawed. The brewery, the world's northernmost, announces its existence with an unmistakable odour of hops. Not to forget the diesel burnt in snowmobiles – it is hard to miss that one unless electric scooters erase it from the scentscape. All that at 78° North, at a place that is said to be warming up faster than anywhere else on the planet, that is undergoing a substantial economic shift from a coal company town to an attractive but disappearing destination, a climate research hub and a green 'testination'. It is neither remote nor untouched, and human presence manifests itself in traces of the past and actions of today so self-evidently that film crews must shoot their landscapes of 'pristine wilderness' carefully, ensuring that no mining or research infrastructure, scooter trails or cruise ships spoil the images.

On a chilly and dark day in November 2018, I was invited to hop on a car in front of UNIS, the University Centre in Svalbard. My youngest son who was 8 months old expressed his mixed feelings about the venture by a choked cry, his baby carrier under my heavy Canada Goose jacket. The guy spoke good English but warned me right away that his working language was Norwegian. 'We won't have a car when we move up here,' I said while trying to climb up to the front seat. 'You won't need it,' said the driver and off we went.

After a few minutes, the car stopped in front of an orange row house with two floors. The 87 square metre apartment we were offered to move into three months later, for NOK 16,661 (approximately €1,600) a month, was situated on the upper floor. The scenery I glimpsed from the entrance was breathtaking. The first week of November still offers a decent portion of twilight around noon at this latitude, even though the sun is no longer to be seen from the end of October and won't appear at the Old Hospital's staircase, *gamle sykehustrappa*, before 8 March. I embraced the mountain at the opposite side of the valley with my eyes and noticed that we would have a lovely view of the church (the world's northernmost, of course) and a strange industrial monument I later learned to call *Taubanesentralen*. It used to be the hub for cable cars bringing coal from several mines around Longyearbyen to town. No longer required for its original use, it is a venue for cultural events with a true genius loci. In spring 2019, we lost the view to the *Taubanesentralen* because of a modular house, one of that year's modest replies to the urgent housing crisis triggered by a complexity of factors, including avalanche and landslide danger, population growth and the state's effort to regain control and re-Norwegianise the town.

The tour round the flat was quick, the previous user was still packing, and we were soon sitting back in the van, which looked shabby compared to the pretentious SUVs I had noticed on the way. Per asked me what I was up to in Longyearbyen, maybe more of courtesy than of serious interest. I replied I was planning to study how people here live with climate change and globalisation. When he said that he moved to the town in the 1980s, I asked a rhetorical question: 'Has Longyearbyen changed since then?' A bitter and sharp reply followed: 'A lot.'

Indeed. The speed of change that only one generation has witnessed in Longyearbyen is overwhelming.

Brief History of Svalbard and Longyearbyen

Svalbard has not always pretended to be a 'miniature of Norway' or 'the world'. In fact, until very late (the end of the sixteenth century) the area was not well documented at all. Russian Pomor hunters might have made use of these lands before that, but it is the Dutch explorer Willem Barentsz who is typically credited as the first historically confirmed observer of the group of islands, with their spectacular pointy mountains, naming the archipelago Spitsbergen. Lacking an Indigenous population, the archipelago triggered awe and fascination and its the raw landscape was an obvious target for colonial imaginaries, as *terra nullius*, a no man's land (and water) full of 'resources'.

Exploration and exploitation went hand in hand in the centuries to come. Whale populations, abundant in the fjords of Spitsbergen, were rapidly devastated by the whaling industry pursued by the Dutch, Danish-Norwegian and British imperial powers. Russian and Norwegian hunting and trapping of seals, walruses, polar foxes and polar bears was also among the economic activities of outsiders coming to the archipelago in the High Arctic, but whaling was the most monstrous, truly industrial venture of Svalbard's early history. 'Whales were butchered and cooked near where they were killed, which is why the remains of whaling stations are found at so many places along the shoreline,' write Hacquebord et al. (2003) in their account of Dutch and English competition in the seventeenth century. In the early nineteenth century, whaling was no longer profitable as the 'stock' decreased dramatically.

More countries engaged in expeditions, driven both by scientific curiosity and hunger for profit. Arctic explorers such as Nordenskiöld, Amundsen or Shackleton passed through on their voyages. In the cultural representations, very much dominated by masculine stereotypes of strength and endurance, there are also other, less mainstream accounts of people meeting the archipelago. One example would be the Austrian painter Christiane Ritter's memories in the book *A Woman in the Polar Night* (2010 [1938]), or the stories of women generally less visible in the Norwegian polar history (Ryan 2022).

The late nineteenth century saw the beginning of tourism in the area, and deposits of black coal were discovered. Mines were opened, not only by Norwegians but also by Swedes, Russians, Americans, English and Scottish (Kruse 2013). Ny-Ålesund started to resemble a settlement after 1916, thrived as a company town until the tragic accident in the mine in 1962, and has since been transformed into a 'centre for global climate research and node of contemporary Arctic geopolitics' (Paglia 2020). The hard extractive industry of coal mining is a powerful component of the identity of places such as Ny-Ålesund or Longyearbyen, founded in 1906. It is in Longyearbyen that the stories I document in this book unfold.

Now represented as a showcase both for the climate crisis and its technological solutions, Longyear City was founded by an American businessman and – as people in town say – a 'cruise tourist' John Munro Longyear. For the first ten years of its existence, it was a multi-ethnic company town created in order to mine coal and ship it south. In 1916, the Norwegian-owned Store Norske Spitsbergen Kulkompani bought the settlement, along with the extensive mining infrastructure, and has been the core stakeholder and guardian of the Norwegian presence since then. Similarly Barentsburg, today seen as a Russian (or more precisely a Russian-speaking) community of about 350 people, was founded in 1920 by the Dutch NESPICO and bought twelve years later by the Soviet Union. Store Norske's equivalent here is the Russian mining company Trust Arktikugol. Together with other Svalbard settlements such as Grumant and Pyramiden, today abandoned or only seasonally revived as tourist sites rather than places where families live, Barentsburg was used as a showcase

Figure 1   Locating Svalbard
Source: Map courtesy of Jakub Žárský.

for Soviet prosperity. There are also numerous differences between the
two largest settlements, but one striking parallel is the initial 'foreign'
investment made within a few years by Norway and Russia (Sokolíčková
et al. 2022), the only two countries that have kept a continuous presence
on the island through financing settlements populated all year round up
to today.

Yet the two Arctic states are not equal in Svalbard. In the work of the
Norwegian historian Thor Bjørn Arlov (2003), the path towards exercising Norwegian sovereignty over the archipelago is outlined, from the
purchase of Longyear City in 1916 up until 1920, when the document by
then known as the Spitsbergen Treaty was drafted, entering into force in
1925. In the same year, the Norwegian authorities also changed the name
of the archipelago from Spitsbergen to Svalbard, keeping the previous
name only for the biggest island, where all the current settlements are

located. While the term 'Svalbard', meaning 'cold edge', is first mentioned in late twelfth-century Icelandic annals (Chekin 2020), with historians disagreeing about which locale the Icelanders actually meant, its embrace in 1925 by Norway was a sign of Norwegianisation – an active construction of the archipelago's political but also cultural identity (Arlov 2020a). Up to now, over 40 countries have signed the Svalbard Treaty, acknowledging Norway's absolute sovereignty over the territory in exchange for equal rights for the signatory parties' nationals in terms of access to the archipelago and the possibility of engaging in economic activities such as hunting or gaining mineral resources. Norway is bound to govern over the territory as 'the best protected wilderness in the world' and ensure peace. Most parties acceded to the treaty in the 1920s and 1930s (among them China, the Soviet Union, the US and the UK), and only a few are recent signatories (such as South Korea in 2012, North Korea and Latvia in 2016, or Turkey's intention to sign, announced as late as 2022). The archipelago is officially part of the Kingdom of Norway but because of the treaty, it is not governed like any other part of the kingdom. Svalbard is not part of the Schengen Area, and many Norwegian laws do not apply here, such as the Immigration Act, the National Insurance Act and other legislation related to social rights and welfare. This is also one of the reasons for the exceptionally low income tax (about 8 per cent, in sharp contrast to the high taxes paid in mainland Norway), which is at the same time the main economic incentive for Norwegians to settle down in Svalbard.

During the Second World War, the settlement of Longyearbyen was destroyed by German forces, but after the war Norway returned and restored it. The Soviet Union also returned. Both countries invested heavily in the mining industry, which provided them with coal and strengthened their foothold on the territory. The second half of the twentieth century was impacted by the Cold War, and the archipelago was a scene of tension and conspiracies, popularly depicted in the movie *Orion's Belt* (1985). Seen from within, to the contrary, people who resided in Svalbard from the 1960s on remember those decades with nostalgia for a time when they were less mobile and less connected to the outer world, but more connected to each other; when the material reality of life in Svalbard was more fitting for the archipelago's location; and when there were no reasons to doubt Longyearbyen's Norwegianness.

In the 1990s, the trend changed, taking a direction inspired by the new order in a suddenly unipolar world, where Russia – unlike the Soviet Union – was no longer perceived as a major threat and globalisation accelerated.

Operating mines were fewer, leading to a decreasing Russian population, in contrast to Longyearbyen, which started to grow and become more diverse and international. Here, the standard of living quickly rose and so did energy and goods consumption, resulting in increasing amounts of waste and pressure on infrastructure. Instant travel has become a simple, cheap and, to many, mundane activity, contributing to speeding up the volume of traffic by both plane and cruise ship. After the turn of the millennium, information technologies and social media made the virtual image of Svalbard widely accessible and tempting. Tourism was booming. Following a thread already to be found in the governmental White Paper from the 1970s (Norwegian Ministry of Justice and Public Security 1974–1975), it was chosen by the Norwegian government as the new economic backbone of Longyearbyen, which was then developing fast both as a science and technology hub, and as a tourist destination. Transnational migrants settling in Svalbard without a visa could live there while staying connected to family, friends or employers scattered worldwide, and Longyearbyen grew bigger, more dense and more complex. Even in the mining

Figure 2   Longyearbyen's coat of arms:
Black sky, with snowy mountain penetrated
by a mining tunnel
Source: https://commons.wikimedia.org/
wiki/File:Coat_of_arms_of_NO_2100_
Longyearbyen.svg

industry spirits were high in the early 2000s, and more Norwegian mines were opened thanks to massive investments. Falling coal prices and rising environmental consciousness contributed to a change in the late 2010s. The newly opened Norwegian mines were closed and are now being left to go 'back to nature' (Avango et al. 2023; Ødegaard 2021, 2022). The last operating mine is Mine 7 (*Gruve 7*), but its life cycle is also coming to an end, with mining activity likely to terminate in 2025, with a few more years to clean up.

For about 115 years, Longyearbyen had a black heart beating in the wild white north to the rhythm of industrialism. Longyearbyen's post-industrial heart is now expected to turn green.

My Travelling Companions: Engaging with Anthropological Theory

In this book, I wish to unpack the paradoxical nature of such a turn in the context of other changes on the move. I peel back the layers of Svalbard's paradox with the help of ethnography, entangled in the questions the place poses through the stories of my interlocutors, myself and my family.

To do so, I engage with anthropological theory that – throughout the process, from writing the first funding application in 2017 until revision of the draft manuscript in 2022 – spoke to the themes emerging from my research and helped me to develop and anchor my argument. Such an engagement also means selecting authors whose work resonates with, or in a constructive way challenges my own thinking, and thus cannot be exhaustive. This book is an outcome of a thinking path, where distinguished companions walked the second mile with me. Those are authors interested in climate change and globalisation, overheating, scale, temporality, extractivism, justice, violence, nationalism, and also the Anthropocene.

My work is inspired by the approach the team of Thomas Hylland Eriksen, my research project's mentor, developed within the Overheating project (Eriksen 2016; Eriksen and Schober 2016; Pijpers and Eriksen 2018; Stensrud and Eriksen 2019). In their introductory essay, Eriksen and Schober (2016: viii) explore the 'intricate meshwork of partial connections […] cultural hybridity, social differentiation and their counter-reactions in the shape of identity politics attempting to reinstate boundaries and purity'. The idea of destabilised identities resonates with what I found in Longyearbyen, very much part of 'a fast changing world with rapidly increasing connectivity and mobility, with mounting environmental challenges, rapid economic transformations and the rise of often virulent nationalisms', where 'forms of belonging to places, groups or communities are being challenged in new ways' (Eriksen and Schober 2016: 1).

Eriksen's work helped me see the concrete connection between two abstract phenomena, climate change and globalisation, which both have to do with overheating intrinsically linked to capitalism. Here also the issue of difference comes into play; difference underpinning a variety of choices and life-ways that can lead to alternative futures, instead of a monolithic linear story in which the future has already been decided. In his editorial commenting on the 2022 IPCC report, Eriksen (2022) sharply criticises the absence of a nuanced anthropological perspective enabling 'a sustained critique of corporate capitalism conspiring with governments' (2022: 1). Dominant narratives (co-created by initiatives such as the IPCC) of how 'the Anthropos' should 'fix' what Anna Tsing (2015b) calls the 'capitalist ruins' disregard what anthropologists have long been insisting upon, namely that 'every event takes place at a particular place and a specific time' (Eriksen 2022: 2). One-size-fit-all solutions are thus risky to pursue, binding us further to the vicious circle causing the problems we are facing. The idea of Svalbard being the world in miniature seems misplaced in this light. Yet it is *part of* the world, and my aim is to document it while paying meticulous attention to the specificities of the place in the time ... of the Anthropocene?

It turned out to be impossible not to refer to Haraway's (2016) 'staying with the trouble'; the Anthropocene lingers in our thinking like climate change and globalisation, becoming just as heavily discussed as identity or culture. There is beauty in the possibility of having a conversation with minds such as those of Latour (2004, 2014), Head (2015), Moore (2015) or Haraway et al. (2016), who put a finger on the ambivalence of the concept. The reasons why I choose not to operate with the term in the book are manifold.

The first pulls back from the nonsensical claim that Svalbard is a miniature of Norway or – worse – the world, which is not a unified place. Longyearbyen is a particular place, enmeshed with numerous processes of many layers, and it blurs the understanding of what is going on if we draw a line from 'unsustainable' to 'sustainable' reducing the path towards sustainability to a technological fix. The narrative of the Anthropocene is linear, modernist, cherishing 'a perfection-yet-to-come' (Haraway et al. 2016: 547): a fairy tale, making the messy world safer and legible, yet tricky if accepted as a political manual. There are more arguments against the

Anthropocene becoming a fairy tale of humankind that realises its mistakes just in time and comes up with a technological solution to the many human-induced crises. One of them is the concept's anthropocentrism, while studying multispecies communication and collaboration might take us further than remaining stuck with the human-species-centred discourse of competition, culture versus nature conflict, and the obsession with resilience, vulnerability, ecosystem services and the like. Multispecies ethnography is not what *The Paradox of Svalbard* offers, but it could have. I chose a different path, but I do not see it as the only possible one.

Some authors, such as Moore (2017, 2018), argue for the Capitalocene instead, rejecting the shallow historicisation of the Anthropocene, in which all people bear equal responsibility for global environmental damage and social injustice. Moore urges paying attention to the issues of profit, power, exploitation of the marginalised, including nature, and the machineries of state, capital and science in their efforts to make everything legible. Such a perspective proves fruitful when unpacking what the abstract beasts of climate change and globalisation mean in the Arctic settlement of Longyearbyen, a place colonised by the Western discourse of modernity, scientism and progress pursued through the reductionist logic of the state.

Yet other influential thinkers, such as Chakrabarty (2017), argue against the Capitalocene:

> Globalization and global warming are no doubt connected phenomena, capitalism itself being central to both. But they are not identical problems. The questions they raise are often related, but the methods by which we define them as problems are, equally often, substantially different. Social scientists, especially friends on the left, sometimes write as though these methodological differences did not matter. (2017: 25)

It would indeed make little sense to claim that climate change and globalisation are the same. Yet they have some things in common, capitalism being one of them. In Svalbard, climate change is an expression you cannot miss when talking to people: scientists, journalists, politicians, young people, retired miners, nature guides – all have climate change in their active vocabulary. The way they understand it, feel about it and see it impacting their lives differs largely, which is also what drove my ethnographic interest in the issue. Globalisation, on the contrary, neither makes it into the headlines of the local newspaper nor does it feature in the rationale of numerous research projects, stories told to the tourists or conversations among the old-timers. If anything, then *geopolitics* would be the word people use to touch upon the complexities of the hyper-mobile, heterogeneous and transient population dependent on employment in spheres where flows of goods, people and capital are key, but where all these processes unfold with Norway's geopolitically driven policy for Svalbard in the background. Unlike other Arctic locales, inhabited by Indigenous peoples who developed place-specific understandings of processes they were part of and cultivated a humble sense of belonging throughout centuries, Svalbard keeps being appropriated on the basis of economic and political interests of 'outsiders'. Appropriation of Svalbard by countries, businesses, discourses and ideologies, its legal backdrop and political enforcement, is a recurring theme also in the context of simmering conflicts in the early 2020s. There are disputes with the European Union (EU) over fishing quotas available in waters around the archipelago, caused by different interpretations of maritime legislation that came into force after the treaty (Hønneland 1998). China challenges Norway's firm standpoint on the country's right to regulate scientific endeavour; Russia has bold plans for the Arctic, regularly raises questions regarding Norway's interpretation of the treaty and insists on maintaining its presence in Svalbard. In the atmosphere of the 'scramble for the poles' (Dodds and Nuttall 2016), there is also the open question of drilling rights and shipping routes when (not if) the ice retreats further.

Svalbard cannot opt out of climate change; it is there and rolling, manifesting both in environmental changes and in the minds and mouths of people. The same goes for globalisation; Svalbard cannot isolate itself, it will stay entangled with living entities (in my work I focus on people) and things (including money resulting in unequal profits) on the move. Both the Anthropocene and Capitalocene (and Chthulucene, Plantationocene and other suggestions I am not even aware of) have their merits and pitfalls, but I present my work without the feeling I must choose another 'theory of everything very fast' (Haraway et al. 2016: 561). Overheating is general enough. The good thing about overheating is that it is an open-ended story; my contribution lies in showing in what sense Longyearbyen in Svalbard is overheated, and also how some of the efforts to cool it down leave out important aspects of sustainability.

Before I describe the backdrop and the logic of my argument in more detail, let me acknowledge how Latour's (2004) understanding of engaged anthropology influenced why and how I frame my work, and what I see as its meaning. In his self-critical appeal to return to empiricism, description and getting closer to facts, he warns against critique running out of steam. Critique is disempowered when we 'believe that there [i]s no efficient way to criticise matters of fact except by moving away from them and directing one's attention towards the conditions that made them possible' (Latour 2004: 231). Instead, he suggests moving back to matters of fact and, further, to matters of concern; that we stop debunking and start caring. Not an easy task; destruction takes less time and effort than construction (not constructivism). I did my best, though, to focus on matters of concern hiding in a single question: *How to live in a warming world where many desire what a few keep for themselves?* Latour, by the way, is one of those who accepts the 'poisonous gift' I am (unsuccessfully?) trying to stay away from, stating that 'the Anthropocene pushes anthropology to the centre stage and requests from it to be worthy of its original mission' (Latour 2014: 139-AAA 8). Such a call is intimidating, but it is also reassuring to read that 'all field studies are studying devastated sites in crisis' (meaning the many crises are real and deserve attention) while 'there is no common world, and yet it has to be composed' (Latour 2014: 139-AAA 12). Globalisation on a planet that is being ecologically ruined does not imply there is a unified world or that things are the same everywhere. It means that things are connected, and through understanding how things are connected *somewhere* we gain reliable material for a comparison with *somewhere else*. At the same time, a case study like this one of Longyearbyen can be used to identify the unspoken and unheard stories that cast a shadow on what, from the outside, might look like a striving for sustainability. The stories I document are unique to Longyearbyen, but they might also be symptomatic of the warming and unjust world. It is the reader's task to judge this.

Obviously Overheated? Scratching the Surface and Building a Scaffolding

At first sight, overheating effects are pronounced at all levels in Longyearbyen. Climate change is fast – it is seen and sensed: higher temperatures, more rain, permafrost thaw, landslides, avalanches, less sea ice, glacier retreat. At the same time, the narrative of climate change is being produced on the island through climate science, and the discourse is setting the agenda for life in Longyearbyen. The economic shift from coal mining to other, softer extractive industries triggers change that has foreseen and unforeseen consequences, and it's the unforeseen ones (or anticipated but not really cared about) that I am interested in. From a 'company town' that was predominantly Norwegian, with a class divide respected by all, but also competitive salaries and loose regulations enjoyed by all, centrally governed by state authorities looking after the strategic settlement in the High Arctic in the Cold War era, Longyearbyen became an internationalised place with a diversified job market, figuring out what it is and what it should be.

Since 2009, the proportion of non-Norwegians living in town rose from 14 per cent to 38 per cent in 2022, accommodating people from over 50 countries, with the three biggest minorities coming from Thailand, Sweden and the Philippines. Tourist and service industries, together with research activities, attract a somewhat different mixture of the new 'locals' compared to the times before everything started to gallop. While some processes, such as the increase of tourism, international migration and climate change urgency with all its complexities have an overheating effect, other trends that manifested with a growing intensity during my fieldwork seemingly cool Longyearbyen down. The Covid-19 pandemic belongs to those unexpected ones, putting a halt to tourism, temporarily silencing the booming vibe of the town. Yet after the paralysing exceptional state of affairs right after the pandemic broke out in spring 2020, the impacts seemed to have had a further overheating effect. Inequalities rooted in the pre-pandemic status quo solidified and became more apparent (Brode-Roger et al. 2022), as did the fragility of the tourism industry, dependent on instant and cheap mobility of people. Also the central government tightened its grip, suggesting adjustments to existing regulation with a potentially cooling effect. The spectrum of laws and rules newly approved or under consideration (as of 2022) range from tourism regulations regarding the fuels used in the cruise industry, movement on and around the archipelago, environmental protection, landing sites and guide certification, to closing doors to families whose children have special needs and to political participation for non-Norwegian residents. Another tool used to cool down and re-Norwegianise Longyearbyen is the housing policy, exercised more through indirect measures, such as increasing state ownership of housing units or renting flats only to those who are likely to fulfil the (geo)political aims for Svalbard. The scene is complex and one stumbles across many layers, contradictions and paradoxes.

The ways in which both Norway's policy makers and people living in Longyearbyen negotiate and navigate through the changes embody the emblematic questions of our time: How to live in a world where one urgency trumps another? Which values are to be cherished, and at what cost? What happens when the paternalist (Pálsson 1996) argument of a 'fragile environment' triumphs in the discourse, to the detriment of human lives that, in the geopolitical perspective, are seen as meaningless? This book can be used as binoculars to look at Svalbard in the High Arctic, which is attracting ever more global attention nowadays given its strategic importance in the immediate future, through the overheating lens. It narrates a story about the conflict between different scales of the striving for sustainability and the driving forces of politics and the economy in the 'global North'. The northernmost settlement, as close to an ordinary town as it can be in the 'uninhabitable' but peopled Svalbard archipelago, has something to say about climate change, globalisation, inequalities and social injustice, agency and dignity, need for continuity and feeling of loss.

The book is divided into three parts: Fluid Environments, Extractive Economies and Disempowered Communities. These mirror the three layers I was originally interested in, namely: how people live with a changing environment; how they experience the shift in the economic strategy for Svalbard; and how both these changes transform the social landscape in town (and vice versa). None of the parts is a summary of 'results'; instead, I trace my thinking with the people I met in Longyearbyen and whose reflections guided me.

In Part I, I first explore what I learned about experiencing change through talking to natural scientists, and especially geologists. I unpack different meanings changing environments have for Longyearbyen residents, and link these to 'the other changes' that many of my interlocutors felt as equally (if not more) pressing than what we have learned to call 'climate change'. Part I ends with an a chapter about the climate change discourse, which is viscous and ready to use in order to push political agendas that my participants struggle to believe are motivated by truly environmental concerns.

Part II explores how the supposedly new and more sustainable industries of science and tourism hold on to their extractivist features. I challenge the black-and-white representation of coal mining in Svalbard, showing how the green future of Svalbard fails to break up with its bed-fellow of coal mining, still following the same 'take out and sell further' logic. I connect the lived experience of the coal miners, feeling redundant in the process of designing a brighter tomorrow, to the struggles of other 'little people' such as tour guides, left out of the discussion of what it means to make tourism more sustainable. The aim here is to include the issue of justice in a debate where political views from the outside tend to override and disregard how things are felt and lived locally. This is a familiar aspect in the Arctic context, but Svalbard seems to be particularly resistant to mobilising for political action, instead of being used as a political tool.

The last part of the book distils my theorising of what living in a place such as Longyearbyen feels like, and what the hurdles are on the way towards a more inclusive and empowered community. Here the paradox lies in the uniqueness of Longyearbyen fertilising diversity, and the suffocating political framework that squeezes the town into a narrow space with little room to manoeuvre, and where frustration, segregation, disempowerment and exploitation grow.

My Path

When I moved to Longyearbyen, I went with the flow of the field. People always talked about change and politics. The dominant narratives that only intensified during my stay were those about a 'vulnerable environment' and *Svalbardpolitikk*, meaning Norwegian political strategy for Svalbard put in practice. These two were not separate; they were rather firmly embracing and feeding into each other. My pre-fieldwork thinking about accelerating change, which drew my attention to Longyearbyen under the impression of reading about overheating, was challenged by a sort of a movement I traced in the gatherings and happenings of Longyearbyen. That movement did not fit the linear idea of change as development or, worse, progress. A 'winding movement' (Ingold 2007) as it were. A disturbing impossibility to 'move on', a constant reiteration of questions that tend to be pushed aside for the sake of a clearer, less messy story that does promise some kind of progress, but silences paths that are part of what Ingold and Vergunst (2008) call a meshwork. I took seriously Moore's (2015) statement that 'even in an era of rapid change, we still need critical analysis of the characterisation of that change and responses to it' (2015: 28). The ancient metaphor of cyclic return, the *ouroboros*, helped me stop fearing I was making a mistake if I kept revisiting the original impulse behind my research; even though my conversation partners were vocal about issues pointing in many different directions, I was looking for how climate change and globalisation were enmeshed.

The Paradox of Svalbard is the outcome of my individual research project carried on from February 2019 until June 2021. I came to Longyearbyen together with my husband Jakub Žárský, and our three sons Josef, Vratislav and Adam, by that time aged 5, 3 and (almost) 1. Jakub is a polar ecologist who studied at UNIS in 2009 and later came back to Svalbard repeatedly for fieldwork, while doing research also in Iceland and Greenland. It was his world I entered. Our marriage needed this merging, I suffered from the separation when he spent long months in the Arctic, and I was jealous about the otherworldly environment I could not relate to in any way. When our partnership was at a crossroads in 2016, we shared the idea of moving to Svalbard with our therapist. He called it 'an escape phantasy' and did not pay much attention to it. But we meant it. In 2018, I received an EU-funded grant hosted by the Department of Social Anthropology at the University of Oslo and mentored by Thomas H. Eriksen. We packed six 220l barrels, mostly filled with children's books, warm clothes and Jakub's gear, and we left Prague, including the painful memories of a dysfunctional relationship behind.

My 'long conversation' (Gudeman and Rivera 1990) with the people of Longyearbyen did not stop after two years when my project ended. We stayed half a year longer, thanks to two new projects I had become involved in. One, funded by the Svalbard Science Forum and hosted by my department in Oslo, focused on a phenomenon I became passionate about once I understood the dynamics of Longyearbyen better: the offspring of non-Norwegian migrants. The second, funded by BELMONT Forum and involving eight institutions in Europe and America, came up with the offer of a postdoc position at the Arctic Centre at the University of Groningen in the Netherlands. The SVALUR project aims at connecting different ways of understanding and remembering a changing environment. Engaged in a project with environmental memory as keyword and Svalbard as locale, I stayed hooked.

During the two and a half years I lived in Longyearbyen, I recorded my meetings with over 220 residents in an audio or written format. I arrived with only basic knowledge of Norwegian, the language of the majority of the population. The first year of my stay was very much dedicated to acquiring proficiency in the vernacular, which was not easy given that no public language courses were available. Only with time did I come to understand the complexity around this – at first sight incomprehensible – lack of service, but I could not give up as I realised that Norwegian is the entrance ticket to spheres one can never access otherwise. In the meantime, I talked to my participants mostly in English, occasionally in German, Italian and Czech, and in rare cases in Russian and Thai, with the help of a paid translator. When I finally could engage in a conversation in Norwegian, I saw there was a boundary I could not cross regardless of my language competence. Knowing Norwegian did not compensate for my non-Norwegianness.

My reservoir of face-to-face meetings consists of 114 Norwegian nationals (7.5 per cent of the total Norwegian population in Longyearbyen) and 93 nationals of other countries (10 per cent of the total non-Norwegian population of Longyearbyen). Becoming more immersed in life in Longyearbyen as the fieldwork proceeded, I actively tried to reach out to people across categories of age, gender, education, country of origin, job and length of stay. My sample is not fully randomised; in the beginning it was easier to get in touch with internationals who spoke good English, worked as researchers and/or raised children in town, as those were the niches in which I spontaneously found myself. After some time it became easier to reach out to people whose worlds hardly ever impinged on my private universe in town.

Because of the framing of my research, with an interest in the question of how people live with rapid changes, it was particularly important for me to listen to people who have lived through the recent dynamic decades in Svalbard themselves. These old-timers are direct representatives of the social memory that enables people to 'negotiate their identity in relation to the past, and a platform from which they can plan for a common future' (Lyons 2010: 24). I met 31 out of 141 residents who had been on the population register for more than twenty years. That means 15 per cent of my participants have a long personal history in Svalbard, compared to the ratio of 6 per cent in the actual population. The longest personal history within a broader family connection to the transient Svalbard was the life story of a participant whose children are the fifth generation of the family to live in Svalbard.

Having said that, I do not wish to claim that the memories, bonds and experiences of people who have been living in Longyearbyen for less that two decades are in any way less worthy. Some 'parallel structures' in town consist almost exclusively of people with a shorter personal Svalbard history, but those are also very much part of today's Longyearbyen. The current annual turnover of people who come and leave is 25 per cent. In my dataset the level reaches – quite accidentally – 24 per cent, meaning that 50 people I met between February 2019 and February 2021 have already moved away (as of June 2021). Interestingly enough, a few of these 50 have already translocated back to Svalbard as they missed it too much, which is a fairly common pattern not captured by the numbers Statistics Norway provides. There are many registered as newcomers who already have a longer history on the island, meaning there is a sort of an interrupted continuity, invisible in population development graphs evidencing Svalbard's transience.

In 2021, according to Statistics Norway there were fewer than 100 full-time year-round workloads (FTEs – full-time equivalents) in the mining industry, more than 510 in the tourism industry (decreased to about 370 after the pandemic), and more than 230 in research and education. In my dataset people working in these industries are represented respectively by approximately 10 per cent per economic pillar: 9 for the mining company, 53 in the tourism and service industry, and 28 in research and education. I also spoke to people employed in institutions representing the state, local public governance and construction industry, to entrepreneurs, artists, journalists, freelancers, pensioners and people on parental leave. I did my best to capture the superdiverse Longyearbyen of the early 2020s.

Anonymisation of my interlocutors, the voices of whom the reader will hear in the book, is partial. When I reproduce quotes or narrate life stories that are, for various reasons, impossible to anonymise, I obtained written consent for how their words and worlds are framed in my writing. In such cases, I reveal both the name and surname of my conversation partner to acknowledge their contribution to my work. When anonymisation was both possible and necessary, so as not to expose my interlocutors to any negative impacts of partaking in the study, I sometimes changed not only their first names but also gender.

Recording or taking notes while talking to people was only one of the methods I used. Taking part in the town's life was an equally important one, and living in Longyearbyen with my family transformed this customary ethnographic method into a means of survival. I consciously took the decision to engage, and to be transparent and public about my work. This was not unproblemtaic and I had some doubts as I gradually became aware of locally sensitive issues that my engagement necessarily touched upon, be it political marginalisation of non-Norwegian residents or the complex landscape of climate change discourse. In this book, reflection over my positionality in the field is only sketched. Interested readers may watch the feature-length documentary *The Visitors* (CZ-SK-NO, 2022, directed by Veronika Lišková) where the fights I picked on my path are represented by the filmmaker observing the engaged observer, myself.

PART I

Fluid Environments

1
Fairy Tales of Change

There are few topics in anthropology as cross-cutting as climate and environmental change today. For good reasons: while the general awareness of something going wrong was raised for example by Rachel Carson's *Silent Spring* (2015 [1962]) as an eye-opener, scientists across disciplines have paid attention to the rapid changes ever since. Anthropology, like any other field of study, has grown in entropy, specialising in innumerable narrow niches. Environmental anthropology and later also the anthropology of climate change (Barnes et al. 2013; Sillitoe 2021) became prominent. Perhaps thanks to the discipline's inherent assumption about things being always complex rather than simple, the shift of attention towards human–environment relations made the specialisation ambiguous. On one side, you could say anthropologists began to concentrate on a topic demarcated like any other. On the other, it also made anthropologists realise what needs to be done is actually to open up, beyond delineated areas of interest: to multispecies ethnography (Tsing 2015b), concepts such as more-than-human sociality (Tsing 2015a), or other-than-scientific epistemologies (Cruikshank 2005). The 'specialisation' was thus simultaneously a process of making anthropology dare to step out of the shadow of a general Anthropos. You never get a chance to talk about 'the world' with a universal human being when on fieldwork; you only meet people inhabiting their environments. They talk in stories, and their stories differ.

The way we frame our stories sometimes resembles fairy tales, talking about change being no exception. When I came to Longyearbyen and started to ask about change, there was a huge diversity in how people narrated their perceptions, but I could also track certain plots repeating themselves, with some narratives being more dominant than others. When trying to make sense out of the subjective portrayals of what is happening in Svalbard's environments, I stumbled over several obstacles.

First, having a conversation about change is difficult without referring to places and temporalities. What we know from natural sciences is space and time, two separate qualities for the sake of a supposedly objective, delocalised evidence. But for my interlocutors, change was impossible to address without relating in a human, experiential way to the fluidity of time, and without remembering places they know. A change *where* compared to *what*, for *whom* and in comparison to *when*? Far from 'objective' yet real.

Second, it is hard to have a warm-hearted relationship with climate change. The impersonal world understood as a globe (Ingold 2021) is warming, but stories we tell about how we experience it get some *calor* (from Latin, meaning heat; the Italian word *calore* that kept coming to my mind means both heat and warmheartedness) when we bring in the embodied instead of repeating the measured and projected. How can anthropology contribute to an understanding of climate dependent 'on establishing a record, which allows science to engage with all that is beyond the reach of our senses and memories' (Simonetti 2019: 4)? It is senses and memories that ethnographers call into play. Is 'climate' too cold for anthropology?

Here we come to the problem of contrasting the global and the local. After all, both climate change and globalisation are still in the subtitle of my book. According to Ingold (2021), thinking about the 'global environment' (or 'global environmental change') is tricky; instead of reintegrating humans into the world, it 'signals the culmination of a process of separation' (2021: 259). My drive to study the abstract phenomenon of climate change in the concrete settlement of Longyearbyen amidst post-industrial transformation was exploring how it is lived. The 'lived experience', as we have gotten used to saying lately. Ingold (2021) detests the idea of a global environment, because it 'is not a lifeworld, it is a world apart from life' (2021: 260). Also Eriksen's (2016) concept of overheating, in which both climate change and globalisation are critically interrogated keywords, unpacks the many contradictions of our world, which is interconnected yet not uniform or united, and where the modernist drive tries (often successfully) to override lifeworlds with a (capitalist) world system.

Let me invite one more travelling companion to come on my fieldwalk, namely Povinelli (2016) and her analysis of geontopower, which she defines as 'a set of discourse, affects, and tactics used in late liberalism to maintain or shape the coming relationship of the distinction between Life and Nonlife' (2016: 4). The promise of late liberalism and settler colonialism criticised by Povinelli is that we can change while staying the same. In other words, that we can do something about the many crises we are facing without changing the ways in which we understand environment, markets or difference. In this way, the narrative of Svalbard turning green being the miniature of the world is a fairy tale *par excellence*. It freely zooms out and in, from global to local, promotes a radical change while staying the same, and plays with the theme of a global climate crisis without having to bother about local injustices. Not a fairly tale I would tell my children.

In this chapter, I start to explore how people in Longyearbyen live with and talk about change when it comes to the environment. It was 'natural' to begin with natural scientists, and especially geologists; since the University Centre in Svalbard (UNIS) has truly become an internationally recognised climate science hub, the people working there are trained in thinking about change and climate, and they have lived experience with Svalbard's environments.

Geo-logic and Scales of Attention

Scale is not to be understood as a matter of measurement, or big versus small [...]. It does not hold that the more you zoom out, and the more the ground you cover, the more you know. It is not a matter of more or less, but of different points of perception. [...] In social worlds, scale cannot be identified as a matter of empirical size or distance; it is a function of knowledge interest and therefore of epistemology. (Hastrup 2013a: 148)

In Longyearbyen, collective memory is short; something might have happened five years ago that only about 35 per cent of the current population remember. As an early career geologist at UNIS commented:

It's like with football teams, right? People forget how much has happened over the past years and they just remember the last few games. That seems to be a similar thing here. The people who have been here longer, I think, have a bit more understanding.

Among the groupings of people I engaged with in the diverse town, two were particular in the way they approached climate change and especially its visibility (Rudiak-Gould 2013): miners and geologists. We shall return to the former group in Chapter 5.

It is in a way understandable that geologists scale change in a different way than others; thinking about 'deep time' (Irvine 2020) and geological eras seemingly make the perspective of human life (or even only parts of it) irrelevant. In the times of a climate crisis for which cultures and economic systems bear unequal responsibility, I was interested in the reflections of scientists working on large timescales, trained to think in (tens and hundreds of) millions of years.

I am sitting in the office of the head of Arctic Geology department at UNIS. Nervous. I feel the knowledge hierarchy on my body signalling the power relation is uneven. Scientists are high on the ladder in Svalbard (Saville 2019b), expert knowledge is valued and cherished as key for climate change mitigation and adaptation (Meyer 2022). I heard people talking about Hanne Christiansen as the 'permafrost queen'. The palace is humble though, not too spacious, full of books, papers and research artefacts. After a brief mutual introduction I ask what kind of change she has noticed during the 17 years she had lived on the island.

> *Hanne*: Being in a place that has a maritime climate in the High Arctic as Svalbard has, you always see a lot of variability [...]. There was this warm winter, this cold winter and so on ... so changes, it's kind of part of life here. It's been like that from the very beginning. [...] I could talk about it for quite a long time, but now I am talking to a social scientist, which is not usual for me, so I'm probably not speaking the right language. But I think changes are ... kind of many different types [...] you have a very changing population, you have a change in legislation, you have a change in all the conditions that control life up here. And then of course, you have a changing climate and that's the thing that we study, that is why we do research because we want to understand climate. [...] That is what is so interesting about being here as well.
>
> *Zdenka*: Yeah, but would you say that the speed of change is accelerating? Do you feel like ...
>
> *Hanne*: You sound a bit like a journalist! That's what they are typically fishing for: 'Oh, can you tell me this dramatic thing?' And actually, I would say: No. Because I am used to that, I'm used to things changing. I see that, we measure things. We learn more about the developing landscape. But that is my profession. So probably I'm not the right person for you to talk to when it comes to understanding dramatic things because I have a different perception.

Hanne taught me a lesson; actually, a few lessons. Not only did her exclamation reacting to my tabloid media logic bring me to a deeper reflec-

tion on the fetishisation of change and its speed, she also verbalised what
I sensed from my previous interviews with geologists: immersed in the
geo-logic, their scales of attention do not match with those of the anthropologist, too focused on the present in a way that Irvine (2020) urges us
to analyse instead of replicate. For them, scale indeed is a 'a matter of
empirical size', while Hastrup and others who study social words think
'perspective' rather than 'measurement'. And it was my task, not the geologists', to understand this and find a language in which I can not only hear
what they say, but also listen *for* its meanings.

Indeed, it was in conversations with geologists that my question about
change was challenged most often; my encouragement to reflect upon
change was perceived as vague and scientifically unsound:

> I mean, how do you observe change? On a diurnal scale, on a seasonal
> scale, on an annual scale or a decadal scale? As far as climate change
> goes, the terminology that we use is specific. So climate, we're talking
> about a longer period of meteorological, like a 30-year, record. So until
> I have 30 years up here doing the same observations, it's hard to say, you
> know, 'I can see climate change.'

Without me interrupting, the scientist, in this case a young associate professor went on:

> But yeah, I mean, you see a lot of change in the landscape, be it from
> the glacier systems or the slopes or the valley bottoms. There's lots of
> change. And that's related to, you know, a complex series of interactions,
> I guess. But generally speaking, I mean, it's a warmer average temper-
> ature now than it was when I first came up. So the annual average has
> increased and that's predominantly seen in the winter months and it's
> even seen a little bit in the summer seasons. So as far as the temper-
> atures go, things have warmed a little bit. [...] But yeah, I mean, you
> know, the same glacier I used to ski on in January 2010 ... now you have
> to hike quite a bit further to ski on that same glacier.

A lot of natural and physical science produced nowadays is still
anchored in believing that science is an objective, apolitical endeavour
of 'unveiling Isis' (Hadot 2006), reading the (hi)story of 'nature' like a
book through extracting samples and conducting measurements (Hastrup
2013b). Are we trying to know 'how things are', or are we trying to

broaden our understanding of something messy and slippery? Geologists, unsurprisingly, are hesitant to approve climate change visibilism; scientific facts and measurements are not just a different perspective than their own observations of a retreating glacier that is harder to ski on. They are reluctant to acknowledge that different ways of knowing could have equal relevance – some rooted in a scientific approach, others deriving from subjective experience of an environment one inhabits. 'Do you want me to wear my scientific hat, or am I supposed to speak as a private person?' was the usual question I was asked in many UNIS offices, to delineate the terrain within which we were supposed to operate during the interview. In the lives of the scientists, though, these ways of knowing inevitably merge (Sokolíčková et al. 2023); personal experience of a changing environment grows richer the more time one spends in Svalbard or the more times one returns.

Compared to researchers working in other disciplines, geologists (though with exceptions) were the least alarmist and the least anthropocentric, used to thinking about planet Earth without humans: 'I am not concerned about this. I mean, I am a geologist. I have been to northeast Greenland and there are tree remnants that are 2.3 million years old, at 82° North. So I know there have been forests in the Arctic with dark periods.' I couldn't help nodding when listening to this deeply anti-humanistic take, but on second thoughts, there was something disturbingly anti-anthropological about it as well. This fairly tale of change according to the geo-logic allows us to imagine the world which existed before life or before humanity, and in the next step allows us also to imagine both leaving the Earth (Ingold 2021) and the Earth without humans. This alienating imaginary is a triumph of the so-called global perspective 'afforded to a being outside the world, [thus] both real and total' (Ingold 2021: 265), while the so-called local perspective of 'beings-in-the-world (that is, ordinary people) is regarded as illusory and incomplete' (2021: 265). Had I accepted this kind of geological wisdom, I could have stopped there. Yet it was just the beginning, just the first stories of change people shared with me.

In addition, there is an obvious kinship between geology and extractive industry, a manifestation of Povinelli's geontopower. We shall return to fossil knowledge networks (Graham 2020) in Chapter 4. But can we play with our imagination when engaging with memories of the past in a way that draws us into the world instead of making us care less? What comes is an intermezzo with a fossilised wave in the sand.

Daydreaming with Dallmann

On the bookshelves in many offices and living rooms in Svalbard, several volumes have their place of honour. The *Geoscience Atlas of Svalbard*, edited by Winfried K. Dallmann, is one of them. Sometimes my husband took the two older boys for day-long trips and they often returned with stones, in which forms of life about 60 million years old were imprinted. We would gather around the fossils and dwell in awe. Here we are, on an island where the tallest tree is a dwarf willow, a *Salix* just about a centimetre tall. Where trees are dwarf-size, the children felt like giants, crawling on their stomachs and laughing about being 'in a wood'. But the fossilised leaves turned into stones to mesmerise us were big; these were fingerprints of creatures that had been tall and monumental, growing on Svalbard in the Paleogene, an era geologists put at some 23 to 65 million years ago.

There are ginkgo trees, conifers like pines and spruces, and cypresses like *Sequoia*, *Metasequoia* and *Taxodium*. Among angiosperms (flowering plants), magnolias, water lilies, katsura trees, lime trees, poplars, birches, willows and many others have been found, most of them having close relatives living today in more temperate climate zones. When

Figure 3   Fossilised leaves
Source: Photo Jakub Žárský.

compared to the present world, it is an amazing fact that deciduous forests covered large parts of Svalbard during much of the Paleogene, even though the archipelago was situated at around 80° N with a four-month polar night already at that time. (Dallmann et al. 2015: 129)

I found myself uneasy when the children asked questions about Svalbard's deep time; I couldn't find a way to connect our discussions about a warming planet with talking about the past of the environment they knew as their playground. 'So why is it bad that it gets warmer? It would be nice if there were trees here!' I had no simple answer.

One day they found fossilised sand ripples, just a gentle pattern on the stone's surface that kept something as ephemeral as a ripple in the sand for eternity. We found it mind-blowing. Things such as sand ripples and leaves falling from paleogenic trees are not something we foresee will stay archived for the far future, and yet they are. 'You see, boys? Just imagine.'

Remaining Hereish

From browsing geological publications like Dallmann's, looking for fossils and listening to geologists working on Svalbard, I came to the conclusion I should stick to the original plan to follow the twists of the *ouroboros* wherever it would take me. Doing ethnography of the present in a fast-changing place forces the mind to stretch to its past and also engage with envisioning its possible futures. Different periods in the past of the place offer beginnings of different fairy tales that I have been told in the field. 'Once upon a time, there were lush forests at 80° N.' There are no human protagonists in this very long one, which is also its key moral. 'Once upon a time, the archipelago was *terra nullius*.' The geopolitical fairy tale about a pristine icy wonderland full of resources began centuries ago and is still unfolding. 'Once upon a time, Longyearbyen was a Norwegian company town.' In this one, the miner turns from a hero to a redundant and environmentally harmful antagonistic complication, but globalisation messes up with the plot and puts Norwegianness in the spotlight. To be continued. 'Once upon a time, this was an Arctic desert.' Where does this story take us, the one about cold and dry Svalbard where you could go around without a raincoat?

Lena Håkansson was associate professor in quaternary geology at UNIS, studying climate and environmental history. When we met in 2019, she had been a permanent resident for four years, but her first encounter with Svalbard took place in 2002; returning to Svalbard is a common mobility pattern among those who eventually settle down there, or just keep coming back.

Lena: I work with climate research. And from that perspective, we also notice a lot of change within the recent years. [...] The thing that is in your face is more of winter warm spells where you have rainfall in the winter. [...] Something like that happens every now and then for a long time because Svalbard is located right where the Gulf Stream continues, right? So you have a very dynamic climate system here. [...] But we've noticed even more of these rainfalls in the winter in recent years.

Zdenka: How do you feel about that?

Lena: I think it's sad. I mean, pretty much all the scientists working with climate research nowadays agree that the warming that we see globally is man-made, and that's something that makes me really worried. [...] I have kids, they are 8 and 11, so thinking about their future and stuff ... But [...] I want to be hopeful because when I'm in a position where I'm working with this kind of research, I also have a big responsibility [...] when it comes to communicating these results in a way that the society can actually use. It's so easy to just produce headlines, because that's how the information flows nowadays. The big headlines sell, but they are not always the best. [...] If people just give up, if you're fed up with these horror stories all the time, people just don't do anything [...] you know? So I feel worried because this change that we've been seeing in recent years is even more visible in the Arctic. [...] And when you have that in your everyday life, it's pretty real in a way.

Accentuating future generations is a leitmotif in Longyearbyen, which turned into a 'family community' only in the 1970s, when more housing units were constructed and the rules of the company town allowed for a family life (Strømmen 2016). The first 60 years of the settlement only saw a few children, and Eva Grøndal, one of my interlocutors born in the 1950s, was raised by her grandmother on the mainland as her Norwegian father and Austrian mother were not allowed to keep her on the island. Introducing family life to Longyearbyen, today with about 400 children of kindergarten and school age (Statistics Norway 2021) and also a few dozens of people who grew up on the island and consider it their only home, turns the attention of the public discourse towards the future, including the one impacted by climate change.

Just about a month after I moved to town, I took part in a climate strike attended by about 150 school kids and UNIS students, together with some other local residents. One of the young speakers was Erik Ekeblad Eggen-fellner, a high school student and one of those who have spent their whole lives on Svalbard.

> We have already lost two lives because of climate change. Now it's enough. [...] I hear my parents talk about how they drove snowmobiles over the fjords in the past. I was born in 2002 and I cannot remember it being possible to drive on the Ice Fjord (*Isfjorden*), despite its name.

In Erik's appeal, there is a notion of a fairy tale about the island's frozen past now being replaced by torrential rain in the middle of the dark season, slush and mud instead of ice and snow. The point of reference here is stories told by his parents, not the archipelago's deep time. Stories he would like to live through himself, and wishes his own offspring could have the same embodied experience with snowmobiling across thick and safe sea ice. Such appeals are based on visibility of change across just one or two generations. They are built on the contrast between what the environment looked like in the stories of the teenager's parents, both expe-rienced guides and small-scale tour operators, and the impact changing weather patterns have today on their way of inhabiting Svalbard – such as crossing the fjord on a boat instead a snowmobile.

My conversation with Lena took place about two months after the climate strike where Erik made his emotional speech. Lena's thinking about Svalbard's deep past was not different from what I heard from other geologists, but she was exceptional in the way she self-negotiated respon-sibility for reaching out.

> I want to write a children's book about climate change in Svalbard. [...] I'm working with the history of the Earth, which means that I'm also working with climate history. So climate change in a longer time per-spective. Normally when we are confronted with information about how the climate is changing, we are getting information from ther-mometers, instrumental data, right? And that only spans the last 160 years. Which is a pretty long time when we're humans and it's several generations, but it's nothing. Which means that it's super helpful for us

[…] to look back in time and and get the longer time perspective. […] People very rarely think about it and it can be an eye-opener when it comes to understanding how small we are as people.

Natural science, delivering knowledge about ecosystems being subject to accelerated change of weather patterns, altered relations and strengthened dependencies, is good at what Lena calls 'instrumental data'. She wanted to communicate with children about climate change in a relational way, bringing in a notion of past and future that is shared, inclusive for humans. Also, when Erik wanted to move the crowd, he did not refer to a table with regularly measured average temperature at Svalbard airport. He evoked the feeling of loss, an emotion speaking to the heart of everybody who has seen the Ice Fjord frozen, be it in reality or in somebody's memory conveyed through a story beginning 'Once upon a time, this was an Arctic desert.'

One aspect of the *ouroboros*'s winding movement is how it makes us revisit the environment's fluidity. Climate change, as I often encountered it in Longyearbyen when reading the news and scientific reports, but also when talking to people, was narrated in two ways, which are two sides of the same coin. The first version is the linear and rather catastrophic tale of the Anthropocene full of 'hockey-stick graphs', which could still have a happy ending as long as 'we' act 'now'. The difficult part begins with efforts to define the 'we'. As Povinelli puts it:

> We can give up trying to find a golden rule for universal inclusion that will avoid local injustices and focus on local problems. […] It is not humans who have exerted such malignant force on the meteorological, geological, and biological dimension of the earth but only some modes of human sociality. Thus we start differentiating one sort of human and its modes of existence from another. But right when we think we have a location – these versus those – our focus must immediately extend over and outward. The global nature of climate change, capital, toxicity, and discursivity immediately demands we look elsewhere than where we are standing. […] As we stretch the local across these seeping transits we need not scale up to the Human or the global, but we cannot remain in the local. We can only remain *hereish*. (Povinelli 2016: 13)

The second version of climate change is the cyclical and cynical, the one about 'natural cycles' in overheating and cooling down, about the globe which was doing just fine without humans in the past as it will in the future. As I will show in the next two chapters, people experience and tell stories about fluid environments. They relate to 'climate change' because there is no escape from it in Longyearbyen, but the interesting part comes when we, together with Povinelli, remain *hereish* and attend to local injustices. Chapter 2 digs deeper into scales of attention and diverging perceptions. In Chapter 3, I shall elaborate on the tension between climate change manifesting itself in environmental processes and climate change as a discourse. That is where the painful entanglement of climate change in Svalbard politics, one of the many layers of the paradox of Svalbard, can be explored. Scaling and shifting perspectives will be one of the book's recurring themes, though clinging on to what Ingold calls an ontology of engagement:

> The local is not a more limited or narrowly focused apprehension than the global, it is one that rests on an altogether different mode of apprehension – one based on practical, perceptual engagement with components of a world that is inhabited or dwelt-in, rather than on the detached, disinterested observation of a world that is merely occupied. [...] From this experiential centre, the attention of those who live there is drawn ever deeper into the world, in the quest for knowledge and understanding. It is through such attentive engagement, entailed in the very process of dwelling, that the world is progressively revealed to the knowledge-seeker. [...] The idea that the 'little community' remains confined within its limited horizons from which 'we' – globally conscious Westerners – have escaped results from a privileging of the global ontology of detachment over the local ontology of engagement. (Ingold 2021: 268–9)

In this book, I represent the 'little community' of the Arctic Longyearbyen, founded and developed upon the ideology of extractivism, disempowered in the management of its environments and excluded from formulating its desired future, as a living paradox. The first layer of its paradoxical features consists of the clashing scales of attention regarding its changing environments. While the natural sciences try hard to say something 'solid' about climate change, in the stories that people can emotionally relate to environments are 'fluid' – becoming real through the bonds we establish with them, through memories of 'the' past and dreams about 'a' future. The next chapter explores how and why changing environments matter for the inhabitants of Longyearbyen.

2

Once Upon a Time – So What?
Why and How Changing
Environments Matter

On Saturday, December 19, 2015, an avalanche crashed down Sukker-toppen, one of the mountains above Longyearbyen. It was late morning – the town was still waking when the mass of snow smashed into the houses. A two-year-old girl and a father died. On February 21, 2017, another avalanche came down the same mountain and destroyed two apartment buildings. During a period of heavy rain in the autumn of 2016, there was a landslide near the graveyard and another affecting the town dog kennel. Time and again, the residents of Longyearbyen have had to leave their homes: when floods warned of mud slides, when torrential rain increased the risk of devastating flows of slush from the mountainsides, when winter winds threatened to send avalanches into town. (Ylvisåker 2022: 7)

On 10 September 2020, during an annual Literature Festival, one of the many cultural events that Longyearbyen enjoys throughout the year, I attended the debate with the editor-in-chief of the local newspaper *Svalbardposten*, Hilde Kristin Røsvik. Røsvik was one of the contributors to the edited volume *Our Frozen Water: Ethical Reflections When the Ice is Melting* (Helgesen et al. 2020), which is a follow-up work, broadening the scale and scope of *The Ice is Melting: Ethics in the Arctic* (Helgesen et al. 2015). Røsvik remembered her astonishment at the parallels she witnessed when conducting interviews with women in Nepal who live in areas heavily impacted by climate change, and claimed that 'here, we also live with climate change every day'. Earlier on the same day, I was interviewed by an American journalist from *WIRED* Magazine who stated that 'climate change was the only story in town'.

While revising the manuscript of this book, I encountered a statement in an anthropology blog on the Savage Minds (2015) website: 'When it comes to climate change, *anthropology is not the only discipline in town*. And because it isn't, anthropologists may not get the last word on which of our knowledges and knowledge practices are useful, or useless, in the wider climate change arena' (emphasis mine). How to make sense of such claims?

In Longyearbyen, it is impossible to ignore that something is happening to the environment, that the landscape and weather patterns people once knew as 'normal' have changed and keep changing. A rupture in the collective memory was the deathly avalanche of 2015 and the strange weather that kept tormenting the town in the subsequent months and years, as described in the bestseller *My World is Melting* (*Verda mi smeltar*), written by a local journalist Line Nagell Ylvisåker and since 2020 already translated into Swedish, German and English. Svalbard is being referred to as a frontline community (Fraser 2019), a canary in the coal mine (Winther 2015) and a climate change hotspot (Goldberg 2011; Anonymous 2022).

For sure anthropology is not the only discipline in town; its scales of attention often differ profoundly from those of other disciplines, which have a much greater say in strategies put forward to tackle the 'global environmental crisis'. During the last thirty years or so, 'natural science research on climate received 770 per cent more funding than social science projects' (Eriksen 2022: 2). Realising this means also recognising that what I discuss in this part of the book, and what my anthropologist colleagues have been discussing in their contributions during the recent several decades, is talking from the margins (with some exceptions, e.g. Orlove et al. 2008). In addition, statements about climate change being the only story Longyearbyen has to tell (as a 'miniature of the world') raise an immediate suspicion that over-representing one story covers up other stories, those unheard, disruptive and paradoxical ones. But why not start from what people talk about, and how they see the changing environments matter in their lives.

The Party Spoiler, Safety and Its Costs

During the two and a half years of my research, I asked for a signed informed consent from about 220 people before turning on the dictaphone. I remember well the third of these. Marte invited me home and gave me a cup of tea, I was freezing as I had had to walk quite far to her place and the temperatures were well below -25°C. There was a polar bear skin hanging on the wall. Marte had been living in town for 17 years already, and in her perception all the changes were galloping ahead without giving people time to take a breath and figure out what is happening to their town. There was a moment of unexpected intimacy when Marte remembered the avalanche and started to cry. On the audio recording, there is a long silence and then Marte's and my sniffling.

On my way back, I realised that my plan to come to do ethnography using English was a naive one. Marte struggled with finding the right words in the foreign language to capture vivid memories of great intensity, making her cry four years after the event. I felt incompetent, unable to listen to the story in the language closest to Marte's heart, in Norwegian aptly called *hjertespråket*, 'heart language' instead of 'mother tongue' (*morsmål*). Also thanks to this embarrassing experience I dedicated much of the first year of my stay to learning the vernacular. When we met again in 2020 for a focus group, we could speak Norwegian.

Marte was not the only person who confirmed that 15 December 2015 was a game changer. In people's formulations of how they felt on that day, safety was a clearly emerging theme:

Clare: I see a change in how safe people feel. I have noticed that people are … how shall I put this? The days before the snow storm that resulted in the avalanche, people were looking forward to this storm. Essentially everyone loved this weather, because it makes you feel alive. It's so majestic, it's so powerful. It's nature and we're here because we love nature … a lot of us, anyway. So we were looking forward to this storm. And then the storm came and it was marvellous and it was wonderful. And that Saturday after the storm we all went out and people were digging out their neighbours who were snowed under…. There was a sort of a party atmosphere. But then we heard that there had been an avalanche. And the party was over.

Despite long periods of changing weather patterns and observations thereof, the process of changing human–environment relations can take an abrupt twist because of a one-time event that shatters the mainstream idea about what kind of world we inhabit; a sort of a great rupture (Sokolíčková et al. 2022).

Clare: It was in the middle of the dark season so it was dark, but then it turned not only dark but also black. And ... everyone pulled together in town, the way we do – we all gather together and we work together and support each other. And we still do that. But I feel that there is a new ... fear of nature. That we can't be as happy about the storms as before. I still love a good storm when the wind takes all of the house and shakes it and I feel, I hear ... it sounds as if someone was thumping up the stairs. I love that. But at the same time I stop and think, hmm, this is serious business.

The serious business of climate change, an abstract phenomenon, revealed itself to the people of Longyearbyen through the concrete event; the avalanche 'gave climate change a face' (Meyer 2022: 5). In Clare's account, there are also other interesting hints; for example the claim about valuing nature as the main reason for living in Longyearbyen (for more context see Salem et al. in prep.), or the issue of exceptional solidarity within the community. We shall return to both in more detail in Part III. The main character here remains the avalanche, the only one with fatalities occurring in town (fatal accidents outside urbanised areas are not too rare in Svalbard).

We have become used to hearing about disasters such as avalanches, but also other life-threatening events associated with climate change; some even call the increased frequency of extreme events 'the new normal' we must adapt to. I prefer to stay away from the term 'adaptation', for similar reasons that I tiptoe around using the term 'Anthropocene'. When we look at the avalanche as a 'matter of concern' (Latour 2014), it teaches us something about the paradoxical features of its impacts. In other words, the paradox of how it matters.

Svalbard is a climate change hotspot according to the many parameters scientifically measured and projected, such as average temperature increase, length of periods with snow cover, sea ice occurrence and increase in precipitation. Without downplaying the sorrow of losing two lives in the party-spoiling storm of December 2015, the impacts of the avalanche cannot be compared to the devastating effects of avalanches and glacial melts in other parts of the world; one example is the Peruvian case described by Carey (2008), where tens of thousands of people were killed. For the people in Longyearbyen, two of their fellow residents died and it was two too many. The impacts of the event are large not only in the pain of the personal loss, which is impossible to quantify, but also in its local afterlife.

As Meyer (2022) describes, the avalanche(s) led to a revised areal plan, identification of risk zones, the demolition of almost 140 housing units (mostly in the Lia neighbourhood), the relocation of the student housing, art gallery and craft centre, closed buildings and roads, repeated evacuations and the building of extensive avalanche barriers on the slope of Sukkertoppen, together with a high wall at the hill's foot. All these measures required substantial financial investments, backed up by the Norwegian state budget. The authorities, especially *Longyearbyen Lokalstyre* (LL, the municipal council) sued by the parents of the two-year-old girl killed in the avalanche and others whose property was destroyed, became nervous; in the eyes of some even hysterical. A resident who had spent 50 years in Svalbard commented:

> In my view, it is pure madness. There have been some geologists up here and then they have found out that there is a danger here and there, but not a single politician asks critical questions. They buy it right away and tear down the whole of Lia [which was] completely unnecessary.

With the discourse of safety, the discourse of responsibility comes along close behind. Is climate change to blame, or can the tragic event be explained by other factors? As Eriksen (2016: 138) puts it, 'the problem is that of interpreting a changing world and positioning it on a moral template of trust, blame and responsibility'. The Norwegian government and also the local authorities choose to represent the disaster as a manifestation of climate change. And:

> if climate change is to blame, political action becomes far more difficult. [...] Shifting blame up to higher and more abstract scales enables corporations and governments to behave more ruthlessly than they might otherwise do, since they can divest themselves of responsibility by referring to, for example, 'the global market' or 'global climate change'. (Eriksen 2016: 141–2)

Some in Longyearbyen seemed quite resistant to this narrative. In 2021 and 2022, I worked as a research contact for the high school in Longyearbyen. For the purpose of her research report, one student interviewed another long-term resident, who had been involved in the avalanche protection system before this competence was transferred to the LL at the turn of the millennium:

> When the Svalbard Advisory Board (*Svalbardrådet*) was transformed into the municipal council (*Longyearbyen Lokalstyre*), the Minister of Justice took with him the cadastre, and handed new papers to the municipal council. Therefore they had no information and everything disappeared, absolutely everything. (Moen 2021: 10)

I did not investigate this; what I know from the old-timers is that the general tolerance of all kinds of risks has changed substantially in Norway (and, with a bit of delay, also in Svalbard) in the course of the past decades, and is currently very low. Thus the changing understanding of safety and how the authorities are responsible for guaranteeing it likely plays an important role in this case (for more context see Meyer 2022). What was clear from the voices I heard in the field was that the afterlife of the avalanche, translated into the focus on safety, was locally contested; it contributed to polarised opinions regarding climate change.

The Svalbard veteran recalling how political structures and institutions changed some 20 years before, and how people coming and going were involved in processes relevant for monitoring and evaluating the local avalanche risk, brings me to an issue of the utmost importance in Longyearbyen; the relation between transience and the memory of the place, including its environmental memory. Without diving into the extensive literature on place and identity, the volume edited by Raymond et al. (2021) offers interesting vistas in scholarship reflecting on what has changed in the world over a quarter of a century since Feld's and Basso's (1996) seminal contribution. In the volume's introduction we read:

> When places change, we open up possibilities for people's senses to change too, while recognising the tension with our yearnings for fixity in an increasingly uncertain world that is facing multiple global challenges. (Raymond et al. 2021: 6)

Longyearbyen has been changing fast, and so has people's sense of the place. The transience of the town's inhabitants causes discontinuities in the place's memory, which influences who shapes its future, and how. The high turnover (about 25 per cent each year) and short average time of residency (somewhere between 4 and 7 years) not only result in a short collective memory regarding hazards, but also make it easier to link them to climate change. In the eyes of many of my interlocutors, transience has prevented the continuation of functional protection measures as well as transfer of knowledge and competence (see also Tiller et al. 2022).

When global challenges such as climate change threaten our ontological security (Manzo et al. 2021), there is a need for an inclusive process of negotiation; to feel safe we might need more (or something other) than a wall several metres high just in front of our doorstep. Disruptions (or ruptures) 'can serve to interrupt an already socially and environmentally damaging way of living and relating to place' (Manzo et al. 2021: 334); they can but they don't necessarily do so. The rupture of the avalanche has ambiguous implications and there are competing narratives about how it matters. In a conversation with a geologist, I encountered a high level of scepticism, yet here pointing in a different direction from the opinion of the resident dissatisfied with the decision to demolish the whole neighbourhood of Lia:

> They realised they had to do something about it because after, you know, decades of people saying, hey, this is going to be dangerous. [...] And it wasn't even until a year and a half later when a couple more homes were destroyed that they actually said, OK, we need to start this. So I am amazed at the the lack of action and support that they've provided in that sense. [...] You can't have people die in their houses and then just like ... argue about whose fault it is, to start pointing fingers and [asking] who's going to pay for it. [...] There's no leadership here, as much as they want to be leaders in this town. [...] So it's good that the NVE [Norwegian Water Resources and Energy Directorate] has taken over.

While the old-timers and experts don't agree on what happens now, they share a notion of something other than a warming climate going wrong in the past. If the 'community' (here meaning the local authorities) concludes it cannot tackle the problem with its own means, externalisation of responsibility, for example shifting it from the local to the national level, is one of the options available. Norway, a leading green nation, has high ambitions when it comes to innovation and climate change mitigation. During a focus group with long-term residents, one of them – a non-Norwegian – complained that existing know-how, used for centuries in Alpine environments for example, is not being utilised in Svalbard:

Norwegians, sorry, often think they know everything best: 'No, we are not going to hear what they do elsewhere, we are going to find out ourselves.' Then they take up a consulting company, they get an invoice, it takes a long time, and in the end it costs 200 million instead of 5. 'Yes, unfortunately, we must take a loan.'

Expert knowledge is highly valued in Longyearbyen, while the disconti-nuities in memory and the remote-control governance regime make those who claim they remember and/or could contribute relevant knowledge feel excluded. Here the importance of identity, attachment and empow-erment becomes obvious. A study conducted in a comparable Norwegian context, yet also with some differences, is that of Amundsen (2015). Amundsen studies the interconnectedness of climate-related happen-ings and socioeconomic change in two coastal communities in Northern Norway. In these small settlements dependent on extraction of natural resources (the fishing industry), tourism is gaining importance in the process of economic diversification. Here too the temperatures are rising and periods with stable snow cover are less reliable. People appreciate nature, the small-scale and safe community, the rich cultural life and good connectivity. After showing that climate change is not the issue that people perceive as the most urgent among the many entangled changes they expe-rience, Amundsen concludes that meanings and values of dwelt-in places are key when facing change that requires some sort of a mobilisation. Her statement that 'places will most likely become more dependent on local initiatives and local strategies' (Amundsen 2015: 257) sounds 'out of place' in Longyearbyen, though, where no strategic moves or investments can be driven bottom-up; Longyearbyen's vision, including the post-avalanche one, emerges *hereish*, put into practice by local authorities but dictated by the state. While, according to Amundsen (2015: 258), 'an emphasis on connection to place and well-being may offer a more fruitful starting point for promoting climate change adaptation than climate change impacts', in Longyearbyen climate change impacts and safety prevail.

Svalbardposten, which is carefully read at Norwegian ministries and received by subscribers mostly in Norway but also worldwide in numbers greatly surpassing the number of local residents, prepared a special English edition in spring 2019 to be distributed among tourists visiting the archipelago during the 2019/20 season. The magazine, with a print-run of 16,000 copies and entitled *Top of the World: Living in the Arctic*, had a headline 'The refreshing chill of Svalbard' on the cover page, featuring two walruses enjoying the midnight sun on an ice floe. Climate change impacts dominated the content of the articles included, next to flattering representations of the community's social life and cohesion. The title of one of the main feature articles in the magazine reads 'We must adapt to a new reality', and lists the main impacts of a changing climate with focus on resilience and adaptation:

> The increase in Svalbard's temperature is now three times as high as that measured in eastern Norway and six times higher than the global temperature increase. [...] One challenge is that making existing buildings secure and new housing will be expensive. [...] We were used to stable cold winters with ice on the fjords and dry summer months. [...] More than half a billion kroner is being spent on new housing and protective measures. [...] Technically we have adapted ourselves and that can provide new opportunities. But how the changing climate affects people here is at least as interesting and at the same time most difficult to say something about. (Rapp 2019: 18–20)

The avalanche as a token of climate change is Janus-faced; it looks back to a sort of 'golden past' when weather patterns were stable, winters cold and summers dry, while it simultaneously looks to a future where weather is a threat and the elements gone wild, with climate change needing to be externally controlled for the sake of feeling safe. In this fairy tale of change, the human protagonist is emplaced nowhere. Key is the ability to come up with large financial investments into smart technological solutions regarding 'the new normal', and a happy ending of lucrative technologies, saleable in other circumpolar regions, might come to pass.

But regardless of the techno-optimistic tone, the spook of the avalanche is still in the air. The feeling of loss, sorrow and altered relations with the environment still pursue those who lived in Longyearbyen in 2015, and connect the stories about the avalanche with other stories about how changing environments matter.

Beyond the Avalanche: Mobility Patterns and the Longing for the Familiar

There are only a few adults who say they are from Svalbard and have no other home elsewhere. One of them is Mari; she has spent a few years here and a few there, but her life path has always led her back to Svalbard, where she grew up as a snowmobile enthusiast, deeply bound to the land-scape and cabin life far away from the ever more disturbing vibe of the 'city' of Longyearbyen.

Mari's accounts of how a once familiar landscape has been rapidly changing and how it depresses her are an example of the solastalgia (Albrecht 2005) I have encountered in conversations with several out of the more than thirty old-timers. The most acute and practical impact of the environment turning hostile to the beloved lifestyle are altered mobility patterns due to less sea ice and shorter duration of snow cover.

Mari: I have a cabin [...]. When I was little, three months a year the con-ditions didn't allow me to get there. Now it's three months a year you actually *can* get there. Because of the sea ice. It's totally upside down.

In the memories of the old-timers, there is a strong focus on the abnormal, the never-experienced-before, the becoming-something-else that comes along with climate change impacts. For those who have spent most of their lives on the archipelago, the changes are continuously visible, year after year, and give little hope that the familiar might return for them to see it. For those who used to live on Svalbard, left a long time ago and come to visit, the changes are fearfully large.

Mari: As kids we used to drive on snowmobiles up to Larsbreen. All the time. It was our playground. In the lunch break we would just whizz up to Larsbreen or Longyearbreen. [...] When people who grew up here come back and see where Larsbreen is now they are shocked. They don't recognise the place any longer. [...] Last winter I went on a boat trip in March. I have never done that before, that was first time in my life – a boat trip in March! There was always ice! It's very sad. It's not the way it should be. The same when people ask: 'When is the best time to come to Svalbard? I'd like to go on a ski trip at the end of April.' Now we don't know what to answer. Or, we do, we say: 'Nothing is normal.' You can't be certain about anything any longer.

Not only long-term residents but also children growing up on Svalbard are sensitive to the unusual weather patterns. The author of *My World is Melting* recollects 25 July 2020, the warmest day ever measured on Svalbard, in an essay capturing the sensory puzzlement of her daughter Lotte:

'We have to go out to feel it,' I say to the kids. 'It is going to be really hot. Maybe warmer than any other day one has ever experienced in Longyearbyen.' My daughter, now 7, opens the door. Her golden hair is whirling in the wind. 'The wind is warm and thick, and it smells differently,' she says, before she runs out in the street with her hands in the air, only wearing her pyjamas. It is 18° now, but the Norwegian Meteorological Institute says it could rise to over 20° later today. [...] Later that Saturday evening [my friend] checks her phone for the latest temperature update. Longyearbyen last 24 hours: Max. temperature 21.7°. It is a new record. The old record was from 1979, 21.3°. The heat continues. For the first time, it is warmer than 20° for four days in a row. (Ylvisåker 2020)

The uncertainty about the future and the perception of the environment as something unfamiliar and threatening has an affective impact on human–environment relations. Unlike in other Arctic locales, where a landscape legible to mental 'maps', walked by generations inhabiting it and using it for fishing, herding or hunting, means a direct threat to livelihoods (as discussed e.g. in Hastrup and Olwig 2012), only a handful of trappers on Svalbard still partially provide for themselves. In terms of food, people in Svalbard are, and traditionally have been, dependent on boats and later also planes bringing provisions from the mainland. In terms of emotions, the interaction with the environment, untamed in the human imagination yet not completely unpredictable, is a strong pull factor for many of those who eventually stay for decades. When I met Astrid, who moved to Svalbard in the 1990s, founded a family and has stayed ever since, she chose to talk about rain in the dark season to make me understand the doubts her family is struggling with:

Astrid: The changes in weather are very unpredictable. All the winters we have had with rain ... like when you are at home and it pours down as if you were in Bergen.... Our worst nightmare is dark season with rain. You know, no sun, rain, dark soil, nothing to do because you can't go outside, four months without the sun.... That does something with you. Then you think, well, this is not the the the place to be. Climate might change how we live here. And if we can't be outside and do what we love, it becomes part of the discussion about whether we stay.

Quinn et al. (2015) argue that a stronger place attachment and place identity also imply more empowerment, which may play a crucial role when facing disquieting changes: 'The extreme events and weather-related disasters associated with climate change can forcibly alter the physical and social characteristics of a place, causing individuals in that location to reconsider their sense of place and attachment' (2015: 164).

Astrid and her family keep reconsidering and fine-tuning their sense of place, emotionally attached to Svalbard where their family history has been unfurling for more than three decades now, playing an important identity-forming role. For those who are more transient, the change is more of a nuisance, a sort of a disappointment one feels when expectations of a white and cold Arctic are not met. As a kindergarten teacher leaving the island after a few years put it, 'When I was moving here, people told me I would never need a waterproof jacket. This was supposed to be an Arctic desert. That's bullshit. It rains a lot.'

Snarling Up in Diverging Stories of Change

Climate change, an abstract and global phenomenon 'somewhere out there' and measured in rigid time scales, is translated into perceptions of changing environments 'right here', the temporalities of which flow according to how we inhabit them. Following the destructive event of the avalanche, there was the subsequent dominant safety discourse and its measures; the solastalgia, depression and disappointment; and alterations in how people inhabit the landscape and how they travel through it – alongside innumerable other stories.

According to a report issued in 2019 by the Norwegian Centre for Climate Services (Hanssen-Bauer et al. 2019), the annual average temperature has increased by 5°C since the 1970s. For the winter, the increase reaches 8°C on average. When I accompanied my children to the kindergarten for the first time in mid-February 2019, and hoped they wouldn't be expected to eat lunch outside in the -24°C cold, a senior teacher (born in the 1950s in Ny-Ålesund, I later learned) mumbled: 'By the way, it should be -35°C now.' When I was given a permafrost thaw sightseeing tour in one of the popular hotels, Mary Ann's Polarrigg, the receptionist showed me a room at the end of the corridor: 'We rent out this one only when the whole hotel is booked. People think they are drunk when they enter; the floor is leaning as the building is sinking into the ground on one side.' In Bjørndalen, my children liked to play at the sea shore close to a cosy cabin, which all of the sudden became less close and, where it used to stand only the cut-off foundation piles reassured us we had remembered correctly. The owners had moved it several tens of metres further from the coast as the ocean was eating up the shoreline. Coastal erosion and sea-level rise threaten all sorts of infrastructure, from cabins to mining remnants, or even just artefacts scattered around the archipelago, protected as cultural heritage and telling silent stories about the past of the place, intertwined with its present (for more context see Kotašková 2022). Just before I left the field, the local authorities decided to move the old cemetery from an area exposed to landslide danger to a safer plot.

I have also encountered many representations of climate change impacts that were difficult to verify. Some would claim the Day of the Sun's Return (*solas tilbakekomst*, traditionally 8 March, when the sun's rays hit the Old Hospital's staircase for the first time after winter) should be moved to 7 March, because the sun comes earlier due to the lower profile of a melting glacier. Others observe earlier arrivals of snow bunting and barnacle geese, or multiple efforts of birds to hatch. All such impressions, regardless of whether they correspond to processes monitored by natural scientists or not, have in common the shared feeling of shifting patterns that co-created the collective environmental memory of the place and its seasonality. The focus on changing seasons and the markers of the flow, such as the sun's or the birds' return, has traditionally been strong on Svalbard, as documented in historical sources such as diaries, letters, travelogues and memoirs. Today, these milestones are abundantly reported on social media, with people frenetically posting images of increasing darkness, or the changing colour of the line above the horizon giving evidence of the ever more favourable sun angle, and first sightings of returning species of birds or marine mammals. The cyclical swapping of the period of polar night with polar day, the spectacular twists in the colour of the light and the (ir)regularities of wildlife determine perceptions of time, including changes in these patterns that humans living on Svalbard use as a tranquillising assurance that the world is still legible.

However, the more notebooks were used up for field notes and the more audio recordings were stored on the cloud, the more I was snarling up in diverging perceptions of change, and the *ouroboros* started to look more like a Medusa with many snakes on her head, not necessarily venomous but certainly confusing. As a way out, I present a few 'figures', with a slight hyperbole called archetypes, that would – with limitations – represent the different framings people in Longyearbyen choose when talking about climate change. When I read Eriksen and Selboe's (2015) study, where they analyse diverging understandings of entangled changes in a small Norwegian rural community, paying attention to aspirations and values and critically interrogating transitioning towards sustainability, my 'figures' could lean against theirs.

The Thai Activist, the Norwegian Miner, and the Swedish Climate Scientist

Eriksen and Selboe (2015) worked in a little Norwegian town in the countryside where 'climate change means very different things to different individuals and groups within a relatively small community and inspires different actions and considerations' (2015: 120). That sounded familiar. In their rural community dependent on farming and tourism, people were coming to terms with rising temperatures and shorter winters. There is obviously no farming in Longyearbyen, but agriculture can be seen as an extractive activity similar to mining if it is profit-driven, enmeshed in market structures and potentially used as a tool for claiming a territory; it can be a way of 'taking something out' of the environment (see Chapter 4). What Eriksen and Selboe (2015) find is that 'among the local population [...] many societal and structural changes are seen as being at least as important as climate change in threatening economic activities, cultural identity, and quality of life' (2015: 123). The locale they studied was going through an economic diversification, towards less farming and more tourism and employment in the public sphere, which is analogous to the situation of Longyearbyen. Their three archetypes emerge as typologies of dreams and values; they generalise dominant narratives of how emplaced people digest the global/local tensions and the changes those trigger. Climate change features here, but the never-ending loops of the winding *ouroboros* keep revisiting other issues people live with as well.

My 'figures' are real people, not sociological constructs. I met them several times throughout the fieldwork and they can be seen as 'stable' in the fluid town of Longyearbyen, though two of them have already moved away since I left. As in the above-mentioned study of a community in rural Norway, choosing these and not others is grounded in the economic landscape of the place. The young adult and student activist is the child of Thai economic migrants who find jobs in the service industry needed for the wheels of tourism to turn. She came in the 2000s and spent most of her life in Longyearbyen. The retired Norwegian man worked in the coal mine on the island in the 1960s, came back later and spent altogether more that 30 years in Svalbard. The Swedish climate scientist first visited Svalbard for fieldwork in the 1980s, has been returning ever since, and lived there permanently for over 10 years. (For a reason, a fourth 'figure' is missing: the precarious international guide. We shall meet them in Chapter 6.) Their scales of attention, the way they negotiate the relationship of the past with the future, and what they see as at stake differ.

Parita explains:

You can see climate change in Longyearbyen very well. When I was little, I remember the ice lying on the fjord. You could go with a snow scooter to the other side. You can't do that any more. And in the summer, the mountains are green. They never have been before. They used to be brownish or yellow but never green. And there used to be white flowers everywhere, on the meadows, plains, mosses. Now they're gone – where they used to be there are houses, hotels.... The town is growing.

This short excerpt from the interview contains several layers of what it entails to live with what the place has been becoming since Parita can remember. She is outspoken about the visibility of climate change, which has to do with the climate change discourse, how it is being locally constructed, received and used (see Chapter 3). Like Erik (see p. 30), she recalls the frozen fjord and the impossibility of crossing it on a snowmobile now as a telling example, and shares how she perceives the colours in the environment have been altered; increased vegetation and the lush green slopes are striking compared to the former ochre shades. Following the flow of her memories, she shifts her attention to the disappearing areas with likely the Svalbard poppy (*Papaver dahlianum*), the mountain avens (*Dryas octopetala*) or the tufted saxifraga (*Saxifraga cespitosa*), which all have white flowers and are common in the town and nearby areas. However, the white flowers are not disappearing from Parita's mental map of Longyearbyen because of climate change; the surfaces on which they used to grow must yield to construction works to build infrastructure that the growing industry of tourism needs to develop. It makes no sense to Parita to speak about environmental change without noticing the 'other' changes that go hand in hand. She is not disrupting a boundary between the world of humans and the world of nature; there has never been any.

Parita continues:

I have heard this thing with the permafrost. That when it melts it becomes methane and that's very dangerous for us. And we know it's melting because if the ground is greener it means that it has melted more. We see that there is less permafrost – it's visible. That's frightening. [...] We shouldn't be able to see it. I am just 18 and I see that the world is going in a direction in which it shouldn't go. I have this image in the head – like a doomsday. Like a picture of a world that is totally destroyed. [...] And if we don't do anything about it, it is going to happen. I am afraid that the future generations will not be able to experience the beauty that we see now.

The urgency of a crisis that haunts Parita in her thoughts was palpable during the interview. Her framing of the visibility of climate change on Svalbard was similar to how many other of my interlocutors used their experience (typically with a glacier retreating year by year,) but the way she linked the visibility (in her case of permafrost degradation) to the feeling of environmental anxiety was unusually strong. In her explanations, Parita used the scientific knowledge available to her, translated into plain words that inevitably lead to inaccuracies. But the intense feeling about the changes she was witnessing energised her and made her engage – in climate strikes, youth organisations, media outreach and beyond. About a month and a half after I spoke to her, I had a tough night when the feeling of weird stickiness and meltdown occupied my resting brain:

I had a climatic dream today. I was walking around in a T-shirt, but I didn't have other shoes than my winter boots and I was so hot, that feeling when its endlessly clammy, it was over 30° and I was thinking, this is not possible, we are so near to the pole and it's so warm. (Field notes, January 2020)

In the same month that I paid a visit to the house of Parita's parents, full of Buddhist artefacts, I also knocked on the door of Bjørn. In his home, religion had no space. He called himself 'an old socialist' and when asked about changes he notices throughout his long personal Svalbard history, he spoke with indignation about the privileged position of people employed by the state, and how people working in tourism, especially the Asian migrants, are being exploited. According to Bjørn, the cause is greed. I encouraged him to trace back to where these changes began and Bjørn mentioned the 1990s, when the Soviet Union, a geopolitical enemy present in Svalbard throughout the twentieth century, fell apart, the state loosened control and the government decided to develop tourism. He saw this direction as a blind alley, and for the first time mentioned the climate cause: 'When you look at the planet it is obvious that to move so many people by plane up here ... [ironic laughter].' He remembered the beginning of the boom as quite slow but then Hurtigruten – the largest tour operating company locally – became involved and the growth accelerated; to Bjørn, the company is 'part of the new world', 'the new industry', where people have to be flexible and there is little possibility to mobilise and stand for one's rights.

I sensed I was encountering a representative of the hard core community of Svalbard veterans, all in one way or another bound to the mining industry thus not exactly climate activists. But surprisingly (or perhaps not at all surprisingly), Bjørn was very attentive to how the landscape, wildlife and weather patterns evolve; he just chose to share first his perception of how the great acceleration of global capitalism, including mobility of people and increasing consumption, impacts Longyearbyen. In his view, economic and societal developments are so tightly interlinked with how we interact with the environment that it was necessary to start from where the story begins: externalisation of costs. Somebody – a group of people, a species, a landscape, etc. – loses but their loss is tabooed, not taken into consideration, or else it is denied as it would interfere with somebody else's gain. According to Bjørn, coal has become unpopular because of the global environmental harm of the product itself, while tourism does not mean despicable products but rather ambiguous secondary impacts in the form of the volume of people who come by boat or plane: 'The emissions have increased dramatically.'

I still had to ask Bjørn about how he felt about the changing environment. He was terrified by the 'doomsday mood' of many and felt it 'destroys the future'. I would not do justice to what Bjørn shared with me if I labelled him as a climate denier; he believed that people play a role in what is happening. But he was hesitant to draw a direct link between human behaviour and heavy rainfall in Svalbard. He mentioned that the ground is being eroded by the seashore but brought up the metaphor of natural cycles. Yet: 'That we are now in a warm period up here, about that there is no doubt. We see that it can be warmer here in the winter than on mainland Norway. That's very peculiar.'

Like other elderly Norwegian men fond of mining I spoke to, Bjørn believed science is important as a channel for technology development.

He was concerned about polluted oceans and had faith in the possibility of cutting emissions. But the change he felt most worried about was the lack of solidarity and shortsightedness of tourism.

During my fieldwork, several journalists, both Norwegian and foreign, approached me when writing their pieces on 'the only story in town', climate change. Not a single one had not yet spoken to Kim Holmén, the International Director of the Norwegian Polar Institute. Kim was a locally and internationally renowned figure whom some would call 'one of the climate prophets', which was a comment hinting at his engagement in conversations and writing ventures with the former Protestant priest in Longyearbyen, Leif Magne Helgesen. Helgesen left Svalbard for his mission in the Middle East shortly before I arrived so the only climate prophet I could meet was Kim.

No doubt Kim got to know important people while working for the important cause worldwide – sharing knowledge about accelerating climate change. When I came to bother him with the film crew that was shooting a documentary about my work and life, the soundman struggled with his long beard while attaching the microphone to his chest under the T-shirt. Kim made a remark about how Sir David Attenborough was 'pushing him' to get him to the right spot in front of the camera, and Kim thought: 'I don't need you.' I knew Kim didn't need me either, but he always accommodated my requests and answered my messages with patience.

When I met him for the first time, in his office full of souvenirs from all around the world, and asked about what he had seen changing, he simply replied: 'All of it.' People, tourism, research, laws and regulations, climate. But he was not a passive observer; he felt his position empowered him to action: 'I have feelings about where we should be going. We can shape the future of Svalbard. I see Svalbard as an important player on the global scene.' Kim talked about people in power who come to Svalbard to see climate change and felt that

> we are listened to so we are very much part of the debate on where the world wants to go, where the world can go and what is happening already that we cannot do anything about. We can teach some of these visitors to think about values and other things than just monetary success.

In the figure of Kim, I met the climate scientist who is aware of the knowledge available but also of the powers that hinder action, including spreading confusion and doubts. Kim believed that to respond with 'hard data – whatever that is' cannot work if the partner responds with feelings.

Kim: How do you, as an individual or as a group, form a coherent picture of the world that you feel secure in? You shut out things. It is attractive to have a simple explanation rather than to weed out this noise of information. [...] We live in a time when people strive for rational arguments. 'We have to find the truth. The science should decide.' Feelings are maybe irrational but they are real. And we have to build a society for human beings.

In the three figures, the local youth activist of Thai origin, the retired Norwegian miner and the world citizen/climate scientist, we see that knowledge and emotions, 'hard data' and feelings, science and embodied experience come side by side. But the way this maze of knowing and understanding is walked differs according to which spot people pick to start the pathway, how they feel while navigating through it, and what they identify as a possible way out. Parita builds on her childhood and coming-of-age memories of the only place she has gotten to know well so far, connects it to what she knows about climate science, gets genuinely interested but also worried, and acts on it best she can. Bjørn looks back to the past and sees the changes in the context of a wider class conflict and fight for justice, now giving way to the *laissez-faire* of capitalism, which so skilfully combines environmental concern with the pursuit of profit at the expense of the less privileged. In his worldview, science also plays a role, but he is critical and looks for disobedient authorities that challenge the climate prophecy. Kim represents it, and prefers the bird's perspective to the meticulously local one. Where Bjørn remembers who stood for which political decision in Longyearbyen 20 years ago, what it meant by then, and what it means now that hardly anybody living in town can remember those events, Kim sees the local pains and struggles as petty squabbles compared to the global cause, in which Svalbard can be used as a spectacular example of climate research hub and an educational tourism destination for future climate ambassadors.

All these voices are present in the public discourse, involved in negotiations of what it is Longyearbyen has to tell 'the world' about 'the refreshing chill of Svalbard' turning unpleasantly clammy. The paradox of Svalbard materialises in the coexistence of climate science pointing at indisputable linear trends, and the non-linear and disruptive perceptions of fluid environments. People shift their scales of attention according to their memories of the past and visions for the future, but also according to their values, identities and senses of place. While there is no doubt Svalbard's climate is getting warmer and wetter, ways in which this local manifestation of a global trend impacts life in Longyearbyen are contested. There is one climate narrative, but many narratives of Svalbard environments.

*   *   *

When trying to figure out how people in Longyearbyen live with the impacts of climate change, my attention was drawn to climate change as a *discourse* – not the line you can draw when checking annual average temperatures, but the many directions in which the accounts of my interlocutors were pointing. This winding path is documented in Chapter 3.

3

The Viscosity of the Climate Change Discourse

Crisis is, in the first instance, an affect-generating idiom, one that seeks to mobilize radical endangerment to foment collective attention and action. [...] It is a predominantly conservative modality, seeking to stabilize an existing structure within a radically contingent world. [...] Crisis and utopia have structured the modernist Euro-American project of social engineering, constituting a future caught between a narrative of collapse and one of constant improvement. (Masco 2015: S65)

Just as Svalbard environments are fluid and we know about them according to both different temporalities by which we fine-tune our scales of attention and different ways of emplacement, talking about climate change is also a flow. In the previous chapter, I visited the perceptions of the avalanche next to perceptions of other environmental changes, showing that the ways in which these changes matter locally are manifold. They span emotional distress and feelings of loss through reconfigured ways of inhabiting the landscape to an intensified focus on safety. These perceptions are intertwined with the people's standpoints (where they stand in the society) and viewpoints (what they see as noteworthy).

What happens when we juxtapose the 'only story in town', which is a rather teleological one, based on taken-for-granted ideas of epochal change, progress and crisis, with the loops of the *ouroboros*? If climate change is a global threat in the times of the Anthropocene, why does it still make sense to look at how events take place at a particular place and a specific time? Composing a common world (Latour 2014) cannot be achieved in a conservative modality; we must dare to engage with politics that goes further and engages in alternative futuring. Before we do so, however, an emplaced understanding of the many stories unfolding in Longyearbyen is necessary. In this chapter, my aspiration is to explore the maze of the climate change discourse as my interlocutors walked me through it as local guides.

Slippery Slopes of Asking about Climate Change

Zdenka: What about climate change?

Terry: As long as we manage to keep the mosquitos out of here, it's quite nice!

After a meeting with another participant with whom I shook hands and sat down at a café table, and who started the conversation remarking 'I hope you won't ask me about climate change', I became nervous. It was not the first time – and not the last time either – that people decided to meet me, despite their doubts as to whether it made sense to discuss climate change with a stranger once again. Instead of solid stairs leading me 'up' to 'better knowledge' about a crisis that occupied my mind both as a phenomenon worth studying and a personal concern, I found myself sliding around on slippery slopes of something that I wasn't sure at all was what I believed.

Almost a year after I moved away from Longyearbyen, a colleague forwarded me a thought-provoking article on solid fluids (Ingold and Simonetti 2022), which intrigued me, and I went on to read a paper thinking with ice and concrete about environmental change (Simonetti and Ingold 2018). Thanks to Ingold's and Simonetti's approach to matter as continuous flux, a story I was puzzled by started to make sense.

Visually, the most spectacular embodiment of the narrative of Svalbard as a climate change hot spot, alongside the verbal one of being 'a minia-ture of the world' inventing climate solutions, is the Svalbard Global Seed Vault. The vault is presented as 'safe, free and long-term storage of seed duplicates from all genebanks and nations participating in the global com-munity's joint effort to ensure the world's future food supply' (seedvault. no). The project, led and funded by Norway, has received several awards. It has been featured in numerous documentaries and news items, and reproduced as a cultural meme in manifold art works. The volume edited by Salazar et al. (2017), *Anthropologies and Futures: Researching Emerging and Uncertain Worlds*, has the vault on the cover.

The vault opened in 2008 but had to close temporarily in 2016 due to severe water intrusion, officially explained by permafrost thaw, higher than average temperatures and heavy rainfall. The event was prominently covered in the national and international media as proof that climate change is threatening humanity, and that the supposedly safe and eternally cold Svalbard is at risk (e.g. Carrington 2017).

In March 2021, Arne Instanes, professor in geo-technology at UNIS, publicly contested this powerful narrative about the Seed Vault by calling it a 'climate myth' (Skjæraasen et al. 2021). The Norwegian national broadcaster reported that the Seed Vault needed a new investment of NOK 193 million and that climate change was to blame. 'But bad planning and construction was the cause of the necessary maintenance after only 10 years, according to experts' (Skjæraasen et al. 2021). While according to the media the news took the Norwegian government by surprise, I was not astonished. My interlocutors often presented a different story, such as the one expressed by Instanes: 'Climate change did not really help the situation but it was certainly not the most important factor. [...] It is a known issue that construction activity causes damage in permafrost, which makes it behave differently than if we had left it in peace.'

This notion of 'better leave it in peace' is identical with what Orlove et al. (2008) mention when discussing perceptions of and relations with moving and sentient glaciers, in comparison to permafrost, the glaciers being even more iconic because more visible tokens of climate change. The slippery slope of permafrost, with a deepening active layer disturbed by construction works building the 'doomsday vault' made of concrete, contaminates the narrative of climate change threatening a project that safeguards the future of humanity. Water runs into the tunnel, and the foundations are less solid than imagined in the icy wonderland on the top of the world. A world full of solid fluids such as ice, permafrost (meaning permanently frozen ground but characterised by an active layer that thaws and solidifies in swings of the seasonal pendulum) or concrete, a world where the discourse of climate change is viscous rather than carved in stone.

As discussed in the previous chapter, talking safety is to talk of climate change's twin on Svalbard. The hesitation and, in some extreme cases, even disgust my interlocutors expressed when suspecting I would ask about climate change, was often accompanied by doubts about misusing the cause in order to pursue other agendas, or to hide human-made mistakes. A geologist speaking:

I don't like how media, and especially the politicians use this climate change as ... I don't know, it's ok to fight for your own thing but to have climate change as an excuse for everything…. Like the houses here were meant to last some 25 years and the foundations are bad so that's really not climate change that destroys the infrastructure. It's human-made. But of course it's not nice to have the feeling that slopes would collapse and more rocks would fall and so on, but on the other hand it's a danger that you just produce in your head. I think it's better to keep your feet on the ground and see what the danger really is, and I don't see that much of it here. It will change and it is important to find out what will change, and build the infrastructure according to that. Rather than spending so much money on removing Svea they should invest in new infrastructure in Longyearbyen.

Svea is a former mining settlement that has been through turbulent times during the twenty-first century and is currently being erased from both Svalbard's surface and underground for NOK 1.6 billion. It is, according to SNSK (Store Norske Spitsbergen Kulkompani – the Norwegian coalmining company in Svalbard), one of Norway's most ambitious environmental projects ever. The case is anthropologically analysed by Vindal Ødegaard (2021, 2022). Ødegaard (2021) calls the developments a Turnerian social drama, and an effort to mark Norwegian presence by human absence. She identifies the 'turning back to nature' as 'part of an environmentalist narrative about Svalbard – one that showcases the archipelago as a venue for environmentally friendly initiatives after mining' (Ødegaard 2022: 2). In her analysis of the performative act through which the new narrative is brought into being, Ødegaard claims that landscape and nature become 'central entities' (2022: 3), and sees this as a move beyond the Latourian culture/nature divide in the modern Constitution. While I agree with most of her reasoning, my experience from the field takes me in a somewhat different direction: through putting 'big' nature and landscape in the centre, I see the 'little' people, and especially some of them, pushed to the margins. Some used to be central but are not any longer (such as the miners; see Chapter 4), some have always been marginal and are increasingly so, to the extent of being centrifuged out of Longyearbyen through the subtle bureaucratic violence of the state (see Chapters 8 and 9, but also Chapter 6). In other words, in my empirical material I found few traces of the narrative depicting 'the archipelago as a site of post-mining utopias, a place where nature can heal and innovative environmental solutions can be found' (Ødegaard 2022: 4) being substantially different from the ping-pong argument jumping between crisis and progress. The anthropologist is not unaware of this:

> The modern narrative of perpetual growth and progress [implies] the imposition of a normative and moral judgment of time. This comes about by displaying a *paradox*, that is, disjuncture between what is or what ought or could be, diagnosing the present, and establishing a certain teleology and telos, a new ultimate aim or objective. Such narratives produce meaning, stories, knowledge, and forms of organization at the same time that they contribute to *silencing other stories*. (Ødegaard 2022: 8, emphasis added)

What I want to achieve with this book is, among other things, to give these other stories a chance to be heard. Svea is an extreme case, though. Longyearbyen has only been through partial demolitions and interventions in the urban landscape, which locals see as dramatic, but they cannot be compared to the shutting down and cleaning up of a whole settlement. But Svea would be a frequent character evoked in my participants' litanies about exorbitant amounts of money being spent by the government on symbolic green gestures instead of solving the troubles Longyearbyen is suffering from locally. And soon the twins of climate change and safety turned out to be triplets; in the stories shared with me, they came along with the issue of control.

Climate Change as Alibi for Control: The Case of Housing

In the representations of the story about the fatal avalanche, be it in the media or in the local collective memory, some segments of Longyearbyen's population are either invisible, or they play a marginal role. While the principle of solidarity is often evoked in interviews where participants remember the event, especially the hours and days immediately after the snow mass hit, ethnic minorities living in town relate to the impacts of the event differently. When being asked about their locally anchored perception of climate change, my Thai and Filipinx participants often shared a general concern without pointing to concrete impacts on their personal lives, with one exception: housing units that were demolished as a safety measure contributed to an even tighter housing market. As Sunny, a Thai woman who was by then expecting soon to be forced out of a flat in a risk zone put it: 'They are chasing us away!'

The economic shift from mining to tourism, research and technological development has led to substantial demographic change (Moe and Jensen 2020), and some residents perceive Longyearbyen, Norway's geopolitical and diplomatic tool at the international level (Hacquebord and Avango 2009), as an increasingly hostile place to live, with existing vulnerabilities of some residents exacerbated during the pandemic (Brode-Roger et al. 2022). Pedersen (2017) presents the case as a dilemma and a security issue for Norway. It is in this context that we can see how housing exposed to elevated avalanche or landslide danger relates to the increase of non-Norwegian population – in the effort to make housing safer, but only for some. I once had a conversation with a Norwegian high school student interested in climate change impacts. 'The government says this place ought to be safe for Norwegian families.' There was a moment of awkwardness when the student gave a second thought to the sentence they'd just uttered. 'Well, all families.' In the interpersonal context of acquainted individuals, to pollute the relationships with an application of the politics of exclusion feels embarrassing.

To understand the riddle of the housing market, it is necessary to draw the legacy of Longyearbyen as company town back into the picture. In Longyearbyen, housing was never meant to be owned privately, it has always been secured through the employer. When I started my fieldwork, the owners of the existing housing units were still relatively diversified. According to a report presented by the local administration to local politicians in June 2019, SNSK owned about 30 per cent of housing units, followed by 14 per cent owned by LL and 13 per cent owned by private persons. Statsbygg (Norwegian Directorate of Public Construction and Property) owned about 11 per cent, and over 30 per cent were owned by other businesses, with LNS (a construction company), UNIS and Hurtigruten Svalbard (the dominant tour operator) as the biggest ones. In 2021, LL and UNIS transferred their housing units to Statsbygg and SNSK, and SNSK also bought the portfolio of Hurtigruten Svalbard. In this way, the housing market, for a short period dispersed among several actors maintaining and renting out their units in ways they found most convenient (e.g. renting to private persons, freelancers, Norwegians and non-Norwegians looking for jobs in Longyearbyen instead of keeping the flat empty until 'the right lodger' pops up), became more easily controlled by these fully state-owned bodies. This move meant that people employed by smaller enterprises that did not own their own housing, and who were neither allowed to build nor able to buy, or people without a locally based employer could not find housing. An acquaintance complained that somebody tried to rent her a minuscule room, a closet without a window (which is illegal) for over NOK 5,000 a month, which made her consider leaving. She resisted and succeeded in finding an acceptable place in private hands, but I knew several people who struggled until they had to give up.

When LL decided to demolish almost 140 housing units due to the risk associated with climate change impacts, parts of the portfolio were replaced by rapidly constructed houses in the neighbourhood of Gruvedalen. Locally, the construction work was met with scepticism as the houses were 2.4 metres taller than originally announced and permitted. While neighbours complained that they seemed to be losing their view over the fjord due to the new buildings, the administration provided Statsbygg (which was building the houses) with an additional permission to build almost one storey higher on the same day the application was received. Monica Mæland, half a year later becoming the Minister of Justice in charge of Svalbard agendas, represented the government while visiting the new block of houses. *Svalbardposten* reported ironically about a conversation unfolding over refreshments served in Gruvedalen:

> The cheerful minister asks the representatives of Statsbygg if this is housing for everyone, and gets the answer 'Well, not quite, no.' You get access to the housing if you work for the Governor, LL or if your employer is part of Statsbygg's housing pool. To fall into the latter category, your workplace must be state-owned. (Solberg 2019)

In this way, adaptation to avalanche risk became entangled with state efforts to (re)gain control over housing. In interviews with people employed in administration, research and/or volunteering in bodies that took part in the rescue action and evacuations after the avalanche in 2015, East Asian migrants were repeatedly mentioned in a specific context. Magnus, a volunteer for a local NGO (non-governmental organisation), remembers:

> As long as people speak English it's ok but if they don't it's a problem. I have seen that when we had the avalanche here. People were hiding [...] when we were coming to get people out of the houses [...] we tried to talk to them and they didn't get it.

Related to the language barrier experienced during evacuations, others reported more people actually living in the flats compared to information provided by the population register. In February 2020, this notion made it to the highest levels of the country's national politics. When the Norwegian Prime Minister Erna Solberg visited Longyearbyen, she expressed the following reasoning regarding what Norway does to increase the ratio of Norwegian residents: 'When it comes to housing policy, we first make large investments related to protection against avalanches and landslides. Then it is important that we have good control over housing construction, in addition to having good control over who lives here' (Wiersen 2020).

In her essay on links between security and growth, and exclusion and waste landing (that is, the dumping of rubbish and waste in poor countries overseas), Tsing (2019) elaborates on how a 'good life' is made possible through the lack of care for a good life for those who do not fit in the mental territory demarcated as ours:

> The good life is built by enclosing spaces of privilege and imagined safety. Construction workers are not invited inside. Environmental hazards are dumped outside. The irony, of course, is that those hazards sneak back in. But that is not the plan. The plan for a good life is to let others suffer for making your security enclosure and for the hazards you produce. (Tsing 2019: 27)

Hazards in Longyearbyen are many, including environmental ones. While protectionism and conservationism are prominent in the environmental regime Norway applies on and around the archipelago, there are heavily polluted sites that the authorities fail to take care of, arguing, among other things, that the cost of remediation is too high. One example is the horse farm close to Svalbard airport, where toxicity of the soil caused by substances used in fire-fighting exercises greatly surpasses levels seen as acceptable by health authorities, and where the owner claims his kidney cancer is not a coincidence (Øystå 2022). Another is a missing water treatment facility, with grey sewage water from the growing town, which is reliant on insufficient infrastructure, flowing straight into the fjord. While university students until recently lived in avalanche risk zones, and migrants working in sectors out of the reach of the state's protective wings felt 'chased away', a good life in safe housing was awaiting the privileged.

The Hostage of Melting Glaciers

> Concerns about global climate change are giving glaciers new meaning
> for many people who may previously have considered them eternally
> frozen, safely distant, and largely inert. Most of the world's glaciers now
> seem to be melting rather than reproducing themselves, becoming a
> new kind of endangered species, a cryospheric weather vane for poten-
> tial natural and social upheaval. Encounters with glaciers during times
> of rapid environmental change produce diverse interpretations. (Cruik-
> shank 2005: 7–8)

Cruikshank's observation on glaciers as endangered species, which, when
encountered, produce varied interpretations, is inspiring in the Svalbard
context. Melting glaciers are truly a character you cannot miss. Not only
do longtime residents remember how far and fast they retreated; in the
case of glaciers, the change is visible within a few years. Even we noticed.
The meltdown of glaciers is a plot that tour guides weave into their storytelling: 'Of course I see changes in glaciers, that is the most common
question of the guests: "The glaciers, are they retreating?" Yes, we can see
that even in this short time.' The guides are saddened by both the process
and their own participation in a business with a significant ecological footprint (more in Chapter 6).

In an article written together with Alexandra Meyer (Meyer and
Sokolíčková forthcoming), we elaborate in ethnographic detail on what
Nuttall (2019) crisply formulates in a generic way: 'In many cases, rather
than being the predominant concern for people, climate change intensifies
the societal, political, economic, legal, institutional and environmental challenges that already affect everyday life in northern communities'
(2019: 65). While a melting glacier is easy to deprive of agency and appropriate as the main protagonist and victim of climate change (which is what
the media often do when reporting on Svalbard), people in Longyearbyen
resent and worry about the retreating glaciers, but they are certainly not
their primary concern. In Parts II and III of the book, I explore several of
these urgent matters that are kept invisible in the streamlined portrayal of
Svalbard as a hopeful miniature of the world.

There is a scientific consensus on Svalbard glacier mass balance being
negative and it is unlikely the trend will change: there might be few
glaciers to encounter in Svalbard by the end of the twenty-first century

(Geyman et al. 2022). Research even indicates that surging glaciers on Svalbard – when they move many times faster than their normal rate – could also be triggered by climate change (Sevestre et al. 2018). It was the surging glaciers, together with thriving wildlife (related to protection against hunting for example) that some in Longyearbyen used as a minor story to counterbalance the sensationalist and alarmist discourse of climate change. They did not necessarily do this because they were living in denial (Norgaard 2011); they rather felt it covers up the complexity of the place and their lifeworlds. In a way, glaciers have become hostage – both those genuinely concerned about climate change and those bothered by the dominance of the discourse use them in their argumentation.

The climate change scepticism of some of the old-timers, most of them living a life story with at least some chapters written when there were working mines in Svalbard, can be understood along the lines of what McCright and Dunlap (2011) see as 'an expression of protecting group identity and justifying a societal system that provides desired benefits'. As put by Mari:

> Of course it's man-made. Many of those [sceptics] come from industry-dependent families. They have worked in the mines all their lives. That's why it's so important for them, they don't want it to be man-made. But we know. Polar bears are full of PCB and toxic stuff. We cannot eat the eider eggs because it's full of poison. I have seen these polluted clouds coming from Asia that the scientists have studied a lot. Snow that melts because of dirt and pollution. But it's natural for those who have spent their working life in the mines ... their hearts .... They don't want to see. They don't want to understand. And when you have scientists who also argue against then these people always choose to believe those.

People feel they, too, have become hostage of a discourse that points the finger: there you see a shrinking glacier, and over there a coal power plant that symbolises our dirty past. Now a greener future awaits us, and who knows, perhaps even some icebergs will be saved for our grandchildren.

The cause of acting upon climate change seemed to be losing support locally by renouncing nuance. It was important for my interlocutor who had been employed in the mine to make sure I understood there are only very few 'cool dudes' (characterised as conservative white males confident they understand the scientific argumentation about climate change but who do not trust it; see McCright and Dunlap 2011): 'Maybe a few of the older ones are totally.... But we have an alternative view on it. [...] Because nowadays it's just like, either you are for or you are against it.' This reductionist image of Svalbard reflects a broader tendency of a 'simplified, mono-story way' (Brode-Roger 2021: 497) of portraying the Arctic as a victim of climate change that needs to be saved (Hanrahan 2017), a story that obscures the complexities of the region and neglects the life stories of the people inhabiting it.

I treat climate change not only in its materiality but also as an idea (Hulme 2009) that is constantly negotiated, and is embedded in politics, values, worldviews, and agendas. Such an approach opens up ways to investigate the viscosity of climate change discourse: considering not just what changes people perceive (observation studies, which might be better termed 'perception studies', as observation draws more on the scientific world of monitoring; see Orlove et al. 2008), but how they receive the climate change discourse and why (reception studies; see Rudiak-Gould 2011, 2014).

What I learned from talking to people in Longyearbyen, including those who think about climate change very differently from me, is that a transition towards a more sustainable society cannot be 'owned' by just a few, and certainly not by those far away from the places where transitions are to be implemented. I would rather agree with Hulme that: 'if we are to understand climate change and if we are to use climate change constructively in our politics, we must first hear and understand these discordant voices, these multifarious human beliefs, values, attitudes, aspirations and behaviours' (2009: xxvi).

According to Milton (1997), 'at the level of perception, truth is not an issue. Perceptions cannot be false; people simply perceive what enters their consciousness. What they make of it, through interpretation, is a different matter' (1997: 486). In order to move towards climate change politics, going beyond symbolic acts and not engaging in exclusive practices of marginalisation, all the economic, social and political entanglements must be taken into consideration. Silencing some voices and (re-)creating myths can hardly be seen as constructive in the painful process of moving towards 'seeing ourselves as part of the very dynamic that visualisations of predictive climate models were describing' (Knox 2015: 95).

*   *   *

The climate paradox of Svalbard lies deeper than in the irony of a climate change hotspot heated up by fossil fuels dug from its own mountains. Through an ethnographic insight into what living with such a paradox means in Longyearbyen, we may unveil new layers. People's objections to replacing one linear narrative with another invite an exploration of the twists and turns of different lifeworlds unfolding during a process of transforming Longyearbyen into an environmental showcase. What co-shapes life in Svalbard is much more than the fluid environments.

PART II

Extractive Economies

4

The Art of Taking Out: From Extracting Coal to Extracting Knowledge and Memories

We spend three days in an old cabin, far away from people and electricity. One of the evenings we plan to go for a swim, naked, as we had done the day before. But a shadow is coming to the bay. It is the cruise ship *Roald Amundsen* of Hurtigruten. 'This is not happening,' I think. Three years ago we came here and another cruise ship landed with 150 tourists on the beach in front of the cabin. The kids felt invaded, they didn't like the cameras pointed on their faces. We explain it to the guides, and say we hope they can land on the other side, at the old trapper's cabin. They did so, and I am so grateful. Later, the whole world got to know that this ship was infected with corona. While I felt it was wrong to enjoy these beautiful summer days like I did, the glaciers near Longyearbyen melted. The rivers rose higher, and the coal mine that provides the local power plant with coal was flooded by the melting glacier on top of the mountain. This is the warmest summer ever measured in Longyearbyen, since records started in 1899. The average temperature was 7.2 degrees, 3 degrees above the normal temperature. The old record was set in 2015 with 6.7 degrees. Today a thin cover of snow is lying on top of the mountains surrounding Longyearbyen. The land is full with expectations of cold and ice. I fear they won't be fulfilled. (Ylvisåker 2020)

The essay written by Line N. Ylvisåker about the warmest day ever on Svalbard since measurements began, from which I shared an excerpt in Chapter 2, continues with her memories of what came next. In this new paragraph, the second protagonist of my research – globalisation – introduces itself, intertwined with climate change. In Part II of this book, I shall attempt to explore the economic paradox of Svalbard, which lies in the *raison d'être* of Norway's sovereignty over the territory: practising the art of taking out while protecting it as 'the best managed wilderness' in the world.

It is not known who first spotted the pointy mountains of Svalbard, and in the north, being first (or being acknowledged as such) has always counted an awful lot. In Longyearbyen, the polar history of Arctic explorers is part of the package. For the colonial mindset of empires 'discovering' new lands and continents, the art of taking something out of the remote and cold places that had to be named and drawn into European maps was the most common way of engaging with them, with little respect for those who had inhabited the lands for millennia. In the case of Svalbard, the exploiters did not have to deal with an Indigenous human population as there was none. During a community dialogue in May 2020, an important figure in local cultural life, Anne Lise Klungseth Sandvik, a former waitress in the miners' canteen who came to Svalbard in the early 1970s and has lived through the accelerated development of the archipelago, remembered a meeting of local and central authorities. It was mentioned that 'it is a blessing for the Norwegian government that there is no Indigenous population in Svalbard'. The perceived blessing lies in the unobstructed ability to rule a vast and strategic territory in the High Arctic where nobody is entitled to claim the right to co-decide on how the place will develop. In the course of a few centuries, what is being taken out of the environment has changed, but has the paradigm changed as well?

Extractivism in Transition?

To extract means to draw out, take out, or copy something out – something one has not produced oneself. Originating in late Latin and gaining its current meaning in the centuries marking the beginning of Svalbard's well documented history of human interventions, the term 'extraction' describes the activities performed at that time just as it does those of the twenty-first century. Recently, critical scholarship has widened the definition of extractivism to 'an analytical and also political concept that enables the examination and articulation of deeper underlying logics of exploitation and subjectification that are central to the present conjuncture of capitalist globalization and neoliberalism' (Junka-Aikio and Cortes-Severino 2017: 177). In the Arctic, the paradox of striving for sustainability while leaning on extractivism is particularly apparent. In this chapter, I will follow the threads of extractivism I encountered in the field, focusing on the economic basis for the settlement's functionality.

The history, and also one of the possible futures of Svalbard, relies on taking something out from the environment; the spectrum goes from whaling, sealing, hunting, fishing or mining in the past to tourism, research and technological development in the present. Some 60 million years ago, when the islands, now known for their barren plains, were damp and forested, large deposits of coal began to form (Dallmann et al. 2015). In the early twentieth century people began 'emptying the bank', as one of my participants put it. The extraction continues, though now directed towards 'mining' knowledge (research), experience and memories (tourism).

Such a pattern – developing tourism and science – is traceable also in other boomtowns (Eriksen 2018) reinventing their identity after the end of natural resource extraction, in some places even side by side. To disentangle a place from extractivism is difficult and Svalbard is no exception, though its struggle is particular in the Arctic context.

> Once the extractivist mind frame has become established in a region it tends to spread and serve as a paradigm for economies and societies. These societies, and in particular certain places and regions, thus enter problematic and hard-to-abandon extractivist trajectories. (Sörlin 2023: 5)

Povinelli's (2016) encouragement to remain *hereish* is helpful when trying to make sense of how the new environmentalist narrative of Svalbard pursued by Norway relates to what happens elsewhere. It certainly is not a place that 'does not count', as Klein (2014) characterises sites of extraction that can be sacrificed, poisoned and destroyed. On the contrary, Svalbard counts more and more, and the narrative of the green transition, already put into practice in Svea (Ødegaard 2022) and being negotiated in Longyearbyen, has taken over the scene. Longyearbyen is expected to transition from an environmental disgrace, polluted and dependent on fossil fuels, to a greening pioneer. From the outside, Svalbard seems to be an example of a paradigmatic turn, but my ethnographic documentation unpacks the paradoxes from within, which – in contrast – allow Svalbard to be framed as a sacrifice zone of a post-carbon utopia, constructed (and financed) by a country heavily dependent on fossil wealth (Adomaitis 2022). The petroleum industry is Norway's most important industry and the country was the eighth largest producer of oil and the third largest producer of gas in the world in 2020.

New kinds of extractive futures are suggested, with tourism and research as central features of the new extractive paradigm. On the other hand, the Arctic coastal states are taking their oil drilling and other extractivist ventures further north – Norway being a pioneer in this regard. (Sörlin 2023: 19)

Norway is a country whose 'entire population, not just workers in the fossil sector or elderly voters, are in some sense bribed or doped with the boosted wealth that extractivism also brings' (Sörlin 2023: 9). It is hard to renounce it altogether (Miller 2021), but some places seem better positioned to experiment with initiatives allowing Norway 'to shine as the world's green do-gooders' (Anker 2020: 240). When the world-renowned Norwegian environmental philosopher Arne Næss proposed science and tourism as the post-mining future of Svalbard during his visit in Longyearbyen in the 1990s, his take was grounded in deep ecology, which opposes anthropocentrism and may lead to anti-humanistic conclusions. In this part of the book, I look at what this shift means for the local human population.

The Paradox of Science and Tourism

To make an analytically sound point about both science and tourism being extractivist kin in the future chosen for Longyearbyen is not trivial. Comparing them is a bit like comparing apples and pears; there are similarities but also crucial differences, which is exactly what makes the comparison intriguing. Saville (2019b: 574) suggests that '[t]he new industries of tourism and research and education represent the "softer" version of extracting value from Svalbard's natural resources.' I followed the traces Saville spotted in the field, and gathered evidence that confirms her findings and goes beyond it, extending the notion of extractivism as related to 'natural resources' to something that permeates also the way people are treated.

While the paradoxical nature of tourism in the High Arctic is perhaps more self-evident than that of science, people's accounts and situations made me ask questions about the similarities between the two. During an interview with a person working in the administration of student services, I was surprised by their critique: 'I am also sceptical towards all those students transported around the island to do research, I feel a lot of it is just a pretext for subsidised tourism of the student elite. Neither do I appreciate that researchers are not forced to collaborate more, there is a lot of research trash lying around when projects are finished.' Some of the UNIS students I talked to were well aware of the paradox of doing climate research in the High Arctic as an accomplice of the aviation industry, one of the biggest CO_2 emitters. But there were only a few who would admit their motivations to come to Svalbard overlap with the motivations of those who visit without a scientific purpose – the tourists – as if leaving behind a significant ecological footprint were better justified by collecting data for a Master's thesis than by fulfilling a once-in-a-lifetime dream to experience the Arctic. In May 2019, when the cruise season began, one of the huge conventional cruise ships, *Aida*, arrived at Longyearbyen Harbour. Thousands of passengers, mostly retired Germans, walked to the town from the port. Those who headed straight towards the UNIS building found a sign on the ground, a white blanket with a teddy polar bear lying on it and, written in German: '*Dein Urlaub schmiltzt 12 m²* *meines Lebensraumes*', that is, 'Your holiday causes the meltdown of 12 m² of my living space.' There was also a footnote on the sign, as in a scientific journal where your statements must be properly referenced, explaining how the calculation was made. The authors were UNIS students who are also often active in the local climate strikes.

People are aware that mobility (be it for leisure, for work, or both) is not a new phenomenon on the archipelago (Viken 2020a) and has a longer history than mining coal; with a peculiar mixture of bitterness and fatalism, some residents comment on the touristic nature of anybody's stay. In 2020, I attended an online seminar organised by Mimir, a Norwegian consulting company for tourism development, to discuss future prospects of Svalbard tourism, severely hit by the interruption in global mobility due to the Covid-19 pandemic. As an illustration of the kinds of experiences contemporary tourists appreciate, 'living like the locals' was mentioned. To my question posed in the chat as to whether examples could be given of what such an approach would entail in Longyearbyen, where 'being local' is such a fuzzy idea, the host answered: 'A great success in Northern Norway, enjoyed by for example Chinese visitors, is the concept of "chaotic breakfast at families with small children" (*kaosfrokost hos småbarnsfamilier*).' I imagined a Chinese couple with a high-tech camera filming my boys' fight over a vanilla yoghurt in our messy kitchen, and getting paid for the peep show.

As people were discouraged from meeting indoors during that time, I went for a walk with Line Ylvisåker and shared with her the promising 'concept' that had left me perplexed. 'How dare they,' she exclaimed. The memories of her family's escape to a bay where they were hoping to be alone but a cruise ship expedition paid a visit faded in importance when compared to inviting tourists to invade your private spaces and consume your everyday struggles as a product.

But still – isn't there a difference between, on the one hand, heavy machinery and determined miners brutally altering the landscape, the seabed or the inner guts of the mountains and, on the other hand, a group of tourists carefully landing with a small boat at a mining cultural heritage site to learn about the past and present from a well-informed guide? Tourism extracts in a soft and apparently clean way, compared to the hard and dirty power that engages with the environment following the logic of 'let's take what is out there before somebody else does it', be it oil, gas, coal or other minerals. But the driving force of 'do it now before it's too late' is present here, too, in overtourism (Saville 2019a), mass tourism (Andersen 2022), and last-chance tourism (Johnston et al., 2012).

In academic literature where resource extraction is discussed together with the booming industry of tourism (e.g. Ruiz-Frau et al. 2015; Sisneros-Kidd et al. 2019), there is often an undisputed distinction made between extractive and non-extractive practices. Others, such as Büscher and Davidov (2016), Byström (2019), Stoddart et al. (2020) or Herva et al. (2020) helped me revisit the issue and critically interrogate ideas about tourism and science being non-extractive. As Revelin (2013) shows in the case of Swedish Lapland, the mining boom appeared almost simultaneously with 'pioneer tourism' stimulated by the romanticised scientific fascination with the remote frozen outpost, an Arctic imaginary well applicable to Svalbard. Stoddart et al. (2019: 8) introduce the terms 'attractive development' and 'experience economy' related to tourism, and claim that 'the rapid and dramatic impacts of climate change on the Arctic underlie the emergence of a global Arctic as an object of scientific and political concern [and] subject to global scientific inquiry and political debate'. Byström (2019) shows, in a case study from northern Sweden, how interrelated resource extraction and tourism are, for example in terms of labour-market processes, or how the infrastructure built to accommodate mining needs also produces access to 'pristine wilderness', a concept heavily used by the state, businesses and media, but also disputed in critical scholarship orbiting around Svalbard. Büscher and Davidov (2016: 161, 166) speak about 'environmental industries', showing 'how the seemingly opposing activities, discourses and political economies of ecological tourism and resource extraction are more intricately entwined than often assumed'.

The figure of the scientist contemplating in the tundra while counting reindeer, drilling holes in ice to take samples, or interviewing participants, seems distant from the colonial mining engineer. Scientific extractivism is delicate. While tourism is dependent on a certain volume, science relies on different financial mechanisms and operates in a different mode. As the volume of scientific activities – despite the recent increase (Norwegian Ministry of Education and Research 2018) – is much less extensive than the volume of tourism, the environmental pressure is minimal. What is more, scientists are typically environmentally conscious people, and care both rationally and emotionally about having the least possible impact in the field.

Here it helps to 'follow the money' and unpack the 'fossil knowledge networks', a term introduced by Graham (2020). He shows how ecologically oriented and publicly funded R&D in Canada relies on the carbon extractive industry and represents 'a means of creating and sustaining narratives and a shared outlook in favor of greening the fossil fuel sector as a "solution" to climate change (as opposed to transitioning away from fossil fuels)' (Graham 2020: 94). Graham also mentions that 'components of ecological science such as conservation and restoration ecology and climatic and atmospheric science, which have grown in the context of the deepening climate crisis, are now also harnessed into carbon extractive development'. As Midgley (2012) shows in a comparison between mining in Svalbard and Nanisivik, Canada, the extractivist paradigm imposed on the Arctic is entangled with what he calls a 'geopolitical economy', in the logic of which capital and the state are co-produced simultaneously (Midgley 2012: 55). Extraction here goes beyond production of 'economically valuable commodities but also produces nature, landscapes, states and the like' (Midgley 2012: 168), as well as – in Svalbard's case – geopolitical presence (Paglia 2020). Production of scientific knowledge is thus a further step in the continuum of resource exploitation in the Arctic, including Svalbard, a place where people often feel their lives serve some larger aims of an economic and geopolitical nature, which is well beyond their control.

As Junka-Aikio and Cortes-Severino (2017: 180) note:

there is nothing natural or self-evident about what kinds of substances, elements, objects, or pieces of knowledge become understood and seen as resources available for extractive operations: the discursive construction of something as a 'resource' always entails the employment of a wide set of knowledges, practices and power relations which regulate how the relationship between nature and the society is imagined and enacted at different points in time and space.

Such a context sheds light on the quotation at the start of the Introduction, about Svalbard being a miniature of Norway and Norway being a miniature of the world. In May 2021, a readers' letter was published in *Svalbardposten*, with the title 'The green shift in Longyearbyen: A castle in the air' (Traa et al. 2022). It was a reserved reaction (written by a think tank consisting of senior white men based on mainland Norway, mostly academics, politicians and entrepreneurs) to another article published a few weeks earlier in the *Stavanger Aftenblad* by Siri Kalvig, board member of UNIS and administrative director of the state-owned Nysnø Klimainvesteringer AS. The article was entitled 'The world's eyes are directed towards Svalbard' (Kalvig 2021) and included statements such as: 'Now a new energy landscape is to be conquered!', 'Longyearbyen is conceptualised as a miniature Norway. A simple community consisting of hardworking pioneers of coal mining and knowledgeable researchers', and 'Perhaps there is a sort of kinship among the coal miner in the north and oil worker in the west?' In the media battlefield, the think tank *Seniortanken* and the standpoint represented by Kalvig seem to clash, claiming they are rooted in profoundly different understandings of extractivism. Seen from within, this is unconvincing.

First and foremost, the recent turn in Svalbard's R&D confirms Stoddart's et al.'s (2020: 18) findings about 'industrial orders of worth – emphasizing scientific and technical innovation and efficiency – [that] are more strongly associated with oil development'. The rhetoric of conquering new energy landscapes draws in a straightforward way on the colonialist mindset, a stylistic move that pushes the button of extractivist lust. Portraying Longyearbyen as a simple community consisting of miners and researchers is misleading in several ways. Not only does it silence the fact that the town of Longyearbyen is much more complex (and complicated) than that, as I will show in Part III. It also depicts people as 'resources' on the glorious path of progress and growing wealth.

Another aspect of scientific extractivism is what people in Svalbard call the lack of 'giving back'. When I switched off the dictaphone in the office of Lena Håkansson, the quaternary geologist who felt responsible for the future, we kept chatting. Only then did Lena share a story that was a game changer in her engagement with those who inhabit areas where scientists come to 'do research'. She was invited to join a fieldwork trip to a mountainous area in Central Asia, as part of a project studying local hydrological and geological conditions. The funding was raised through a grant scheme emphasising the importance of 'translating' the research results into practical knowledge for people who live in the area and for whom access to clean water is of fundamental importance. The research group had been returning to the area for over 10 years by the time of the field trip in which Lena took part. On the way to the field site, the group of researchers had to stop where a recent rock fall blocked the dust road with tons of big boulders. That moment was the first encounter of the scientists with the local inhabitants who – going in the same direction and also stuck on the road – shared they had been curious what those strangers had been doing in the area of their water resource for all those years. Lena was shocked and understood that this kind of extractive and 'accidental' science is not something she would like to be involved in. She was also one of those who supported more interaction between science and tourism in Svalbard for mutual benefit.

> *Lena*: I've been talking to people working as guides here. The guides are the ones meeting the guests who are coming up here, right? And many people who would come and visit Svalbard will want to know something about nature. And here is UNIS. We're doing research about nature. [...] Some of [the guides] take the information they communicate very seriously, but most of them don't. So there are very different stories that circulate, some correct, some not. [...] And this is something I know many of the guides really want to change. But it's very difficult for them to access the information. I would really hope for UNIS to take the responsibility in the local community. [...] It would be super cool to get some kind of a forum, like a meeting place, where guides and researchers could meet on an equal basis because this is what most often goes wrong. [I]f you go to lecture, I will stand up and [...] talk about glaciers and climate change [...]. It will be a very unequal setup because it's me informing you, and it's my words which are not your words. So it's difficult for you to take my stories and make them your own. If we can sit down and share our stories! [...] I'm trying to work together with some local guides to get this started.

To my knowledge, Lena's idea has not been followed up. While in other Arctic locales, collaborations and synergies among scientists and those involved in tourism (be it tour operators, guides or tourists) are supported, for example through citizen science, on Svalbard, tourism and research are supposed to be separate from each other.

In 2021, the Norwegian government started to tighten up the legal landscape of Svalbard through stricter, mostly environmental regulations, making science more logistically difficult and expensive, and also strangling tourism. At a conference in 2022, I talked to a representative of one of the stakeholders involved in the decision-making process regarding environmental protection. I shared my excitement about joining a Dutch scientific expedition to Edgeøya, a less frequently visited island of the Svalbard archipelago, scheduled for July 2022, where scientists were going to be on board together with politicians, journalists and tourists. My conversation partner commented: 'The government does not like this. They are suspicious. Mixing tourism and science is not welcome.'

Throughout my fieldwork, the public debate about the so-called societal relevance of science was high on the agenda. 'Science for Society' was the title of a plenary session at the Svalbard Science Conference, held in November 2021 in Oslo, and I was responsible for co-organising it as a member of the scientific committee. We had some good discussions, but the 'two cultures' (Snow 2012 [1959]) still hung in the air. One of the participants commented in the evaluation form: 'The Science for Society was really bad. That had nothing to do with what the society needs, it was only a session for social scientists to complain that they don't feel included.' Despite the bitterness that people working outside natural sciences feel in face of the political decision not to foster social science at UNIS, and to directly ban it (with a few exceptions such as research on cultural heritage) in Ny-Ålesund, there is a rather shared call for a shift, dropping the one-way communication in which science 'presents the results' and society 'acts on it'. This shift is grounded in an agreement that society has a say in what, how and why is being researched. It was not possible, however, to discuss the implications of science in Svalbard being deeply political and value-laden at the conference. 'Giving back' to the local communities stayed at the level of dissemination and 'transfer of knowledge' that can be applied in, for example, adaptation to climate change.

In June 2021, the Svalbard Social Science Initiative co-organised a public debate about the role of Svalbard church as both a religious and public space. One participant raised the issue that while rich scientific contributions are being produced at UNIS, little of this actually trickles down. While taking notes during the meeting, I realised how often I had heard people lamenting that researchers fly up to Svalbard, do their field-work, 'collect data' and leave to analyse them in distant laboratories, publish in journals nobody reads locally, and the knowledge disappears from the island, unnoticed, not used, not shared. Lectures open to the public at UNIS are mostly in the format of frontal presentations with some time for questions at the end, but the space of science, locally so extraor-dinarily developed, fails to be an inclusive public arena.

The paradox of tourism and science on Svalbard has multiple layers. On the one hand, they are kept separate in the official discourse, and spon-taneous efforts to look for synergies or cultivate existing bonds between the two are discouraged. On the other hand, both tourism and science are being represented as part of Svalbard's sustainable turn away from the fossil past to the green future, but both are embedded in the extractiv-ist logic of taking something out, redefining what can be turned into a 'resource'.

The Labour Day Intermezzo

When the local priest Siv Limstrand asked me to give a speech on 1 May 2021, she wanted me to represent Norsk Folkehjelp Svalbard (Norwegian People's Aid), a local branch of an NGO with a major impact in mainland Norway but also active internationally. The NGO arrived in Longyear-byen in 2019 after local members of LO (the Norwegian Confederation of Trade Unions) reached out and asked for assistance to create an inclusive arena for people who could benefit from facilitated integration measures. Through word of mouth I got to know about the first meeting, joined and volunteered for the organisation until the end of my fieldwork, arrang-ing Language Café meetings and other small-scale events in the public library. People attending our events were mostly researchers and tourism workers, and freelancers (editors, journalists and other writers, graphic designers and the like).

The priest encouraged me to give the speech in English, well aware of the growing language barrier in town (see Chapter 8). When I was prepar-ing for the event, I decided to build my short talk around continuities and disruptions in what values we cherish and whom we acknowledge as per-forming constitutive labour crucial for the town's identity. In the central public space of Longyearbyen, *torget*, the sculpture of *The Miner* (*Gruvebus*) by Kristian Kvakland is perceived as a 'focal point', a stage where many scenes important for the town's life take place. Even though there have also historically been miners hired by SNSK from countries other than Norway (e.g. Denmark or Poland) and there are non-Norwegian nationals among the current team of miners operating in the remaining *Gruve 7*, the industry and the people involved in it are predominantly Norwegian. The company symbolises an incontestable link between the former company town and the Kingdom of Norway.

Dwelling on the paradoxical nature of tourism and science on Svalbard, I made an attempt to highlight the similarities between miners and people working in the extractive industries dominating in the present. Especially when comparing interviews with miners and guides, I noticed that both groups were highly sensitive to inequalities, but while the miners fought their fight in the twentieth century, and by the time of my fieldwork belonged to the privileged (with solid salaries, heavily subsidised housing, full health and social service coverage and numerous work benefits), the guides were suffering from precarious living and working conditions and struggled to unite. I saw extractivism as what co-shapes the lifeworlds of both groupings, but I also wanted to point out it is not the workers who are to blame; quite the contrary, the lives of the miners unfolding in Svalbard entailed in the extractivist mode of existence deserve just as much dignity as the lives of others. It was only an intermezzo, and the figure of *The Miner* remains a lonely one at *torget*, without the company of a figure of *The Scientist, The Guide* or *The Cleaner*.

*   *   *

The advance of industrial capitalism fostered a massive acceleration of extraction, and [...] capitalism and extractivism do rise and fall together to some extent. [...] There is a fundamental incompatibility between prioritizing profit and prioritizing environmental stewardship. [...] To decarbonize we will have to extract, with an eye to reducing extraction wherever possible, mitigating its harms, and distributing the harms and the rewards equitably. [...] The challenge of building a real 'social alternative' to accompany a transition to decarbonized energy remains the major political challenge for environmentalists today. (Miller 2021: 202–3)

The point Miller (2021) is making in *Extractive Ecologies* is disturbingly acute. What needs to be done is to imagine extraction beyond extractivism. Even if we detach ourselves from fossil fuels and radically reduce our use of energy, many materials and minerals will still be needed for renewable energy to be fairly produced and distributed. Also tourism and science may walk more sustainable paths, getting rid of the extractivist logic. Are we living in the Anthropocene, or Capitalocene? Might a hopeful phase of the Anthropocene emerge once its destructive chapter of the Capitalocene is over? Is it possible to think extraction *in Svalbard* beyond extractivism, or is too late now that the new extractivist paradigm has become dominant, supported by oil wealth that can even afford an environmentalist narrative? In Chapter 5, I juxtapose different scales of attention when it comes to 'big powers' and 'little people' encountering each other in times of economic change in Svalbard.

5

Big Powers and Little People: Scaling Economic Change

Throughout the twentieth century, Longyearbyen was a proud company town inhabited by miners and gradually also by their families. During the 1970s and 1980s, the Cold War was on and the Russian settlements of Barentsburg and Pyramiden were still lively, both of them. These decades are remembered by some as the Golden Era of the town's population being indisputably Norwegian, relying on secure jobs in the fossil industry subsidised by the central government. There certainly were some FiFos (fly in-fly out workers) in those times as well, people who would take advantage of the low Svalbard income tax of just a few per cent, work on the island but commute to the mainland whenever possible. For others, who were increasing in numbers, it was more attractive to live in Longyearbyen for real.

But the town never was just like an ordinary mining town in mainland Norway. In 1916–17, the first Norwegian mining companies operating in Svalbard were founded. None was formally run by the state but some high-ranked politicians were shareholders in one of them, Store Norske Spitsbergen Kulkompani. After the end of the First World War, when powerful European states needed some time to recover from the conflict, diplomats of the young, small and rather poor Scandinavian state succeeded in getting the Svalbard Treaty signed in 1920, securing the sovereignty of Norway over the archipelago. The most notable historian specialising in the history of Svalbard, Thor Bjørn Arlov, calls the period of 1920–1925 'Norwegianisation' (*fornorsking*; Arlov, 2020b); during these years, Norway made substantial investments in supporting the country's claims and interests in the area. The United States of America signed the Treaty in 1924, China (Republic of China) in 1925 and Russia (Soviet Union) in 1934. Throughout the whole century since the era of Norwegianisation, it was never in a straightforward way economically convenient to govern over Svalbard. The strategic archipelago has always had a financial cost for Norway.

In this context, it is not trivial to say who the giant is and who the mouse. As I will show throughout this chapter, people who live in Longyearbyen often bitterly comment on the irrelevance of their needs and concerns when it comes to the central authorities making decisions that will determine their future individual lives on the island. If we zoom out a bit more, Norway at first sight appears as weak in comparison with the USA, Russia or China. Weakness is not what characterises today's Norway, however; it is a wealthy and in a way also a powerful periphery nation (Anker 2020). The government is drafting measures at the desk office in downtown Oslo, trying to secure the nation state's interests in the context of global politics. But the context has changed, too. Skipping the horrendous intermezzo of the Second World War, since it does not play a major role in the point I am trying to make, from 1920/25 until the beginning of 1990s, Norway was acting in a geopolitical theatre with a legible script and well-portrayed characters. When history started to end (Fukuyama 1992) and then – not very surprisingly – eventually carried on in the 1990s, things speeded up and became more complex.

Who is Black and Who is Green, Who is Dirty and Who is Clean? From Local Heroes to Necessary Evil

In 1989, the Iron Curtain that once seemed eternal was pulled down and things that 'before' were unthinkable became the reality. This applied also in Longyearbyen, built up around the heroic effort of Norwegian miners. Extractive industry can easily be associated with something heavy – heavy machines, heavy work, heavy visual impact, heavy argument for claiming the territory. Since Article 9 of Svalbard Treaty prohibits naval bases and fortifications, and also any use of Svalbard for war-like purposes, the cumbersome extractive industry of coal mining substituted for 'hard power'. Taking the efficient power bank of high-quality black coal out of the impressive mountains in Svalbard is dirty in two ways.

There is the intimate dirt that the miners had to get rid of in *lompensenteret*, a building now in the centre of the town serving the purpose of a humble shopping mall, earlier situated at its outskirts and used as a public bath, a point where the miners could have their filthy clothes changed and washed.

But there is also another facet of dirtiness in coal mining; a more symbolic and global one. One that clashes with the importance of mining in Longyearbyen, in the individual life stories of miners who together co-created a unique, harsh but also caring community, and who lived their ordinary lives in an extraordinary place and worried about things other than Norway's sovereignty claims – for example health care, food or holidays. These life stories of industrial glory seem dirty now, when we know how detrimental it is for the environment to burn coal.

I am sitting on a cosy sofa, enjoying a pitch black Saturday afternoon in December 2019. We were invited to a friend's place, the kids are playing in a surprisingly quiet way so I have a few minutes to chat with another invitee, a miner I am acquainted with. We start talking about the role mining is playing in today's Longyearbyen, approximately where our last conversation ended some weeks ago. Henrik narrates an anecdote, a story a friend of his told him. The guy was on the mainland, in a city park, and saw a group of environmental activists protesting against something. He approached them and was asked about his job. 'I work as a coal miner in Svalbard.' Zac! A slap in his face followed. I laugh briefly out of surprise. The era of pride for Svalbard's coal miners is over, the era of blame and shame has come. Not yet in Longyearbyen; the town still is a safe and hate-free zone for the local heroes. But their presence is fading. SNSK changed its name to just Store Norske in 2020 (still great, still Norwegian, but not obsessed with coal any more) and decreased the number of employees from the once usual 400 to about 130 in March 2020 (out of which only about 40 indeed mine coal), expecting the numbers to diminish even more and the last operating mine to be closed no later than 2025.[1] The dominant industry and its people are becoming marginal. To understand what it means, Henrik's story will serve as an example.

He came to Svalbard for adventure. At first he just had some random jobs in Longyearbyen and Svea for a couple of years, before he became a miner himself. He kept commuting (by plane, there is no road) to Svea

---

1.   In 2020, when this chapter was first drafted, the expected closure of the last mine was to happen in 2038 at the latest. That is also what my interlocutors working in the mine believed. The decision to close very much sooner instead came in 2021. When Russia invaded Ukraine in February 2022, the discourse changed again, prolonging extraction of coal until 2025 for industrial purposes in Europe. Switching from local coal to imported diesel as a preliminary source of energy until a renewable source is secured is planned for autumn 2023. The reason for this ambiguous decision is the poor technical condition of the local coal power plant, which needs large investments for maintenance.

until 2015, meaning he would see his family only when off work. Then he was offered employment at Mine 7, about 20 minutes drive from Long-yearbyen, where he has been working ever since.

During the first years of Henrik's stay, he saw the mining commu-nity as placed in the centre of everything. 'Of course we were very happy about that. We weren't just something that needed to be here for it all to be working. The focus was on the miners and everything else was built around supporting them and making them do the work as well as possible.' Henrik understands how important it is, in the segregated town, to keep track of what is happening in the other 'bubbles'. Already, from the begin-ning of his Svalbard life, he was connected to UNIS as well – his very first seasonal job in Svalbard in fact was being a fieldwork assistant for a PhD student at UNIS. Without me bringing up the issue of identity, he sponta-neously comments: 'I do not see myself as an archetype of a miner. I sort of have … you can say, more legs in different places.' Mining was not a choice of his heart; he just needed a good job that would make his dream about adventure in the Arctic come true. 'I am quite proud of being a miner but […] it is not an identity for me, like it is for many other people.'

In the 2000s, 50 new positions were open for miners in Svea. Henrik got one of the jobs, not because of qualifications since he had none, but because he was already in Svea, the guys knew him and he knew the place. The company trained him and his mining career began. This was the period of 7-day shifts, meaning that some would work for a week in Svea and then leave for the mainland, while others would go to Longyearbyen for a week. 'It was quite a tough life at that time because it was basically 7 days working in Svea and then […] 7 days here in town, maybe 5 of them spent at Karlsberger.' Karlsberger Pub, or [ko-be] as people call it, pro-nouncing the initials KB the Norwegian way, is a cosy bar situated opposite the grocery store, with loads of all sorts of alcohol. It tries to clutch at the image of a miners' pub also in 2022, with rusty artefacts displayed around the place and touching black-and-white, large-format portraits of miners on the walls. 'So people were burning the candle on both ends,' says Henrik and chuckles. But he also recollects how high people's spirits were: every-body knew each other, work comrades were their closest friends. Robert Hermansen, the boss of the mining company, was at that time engaged in supporting small communities in Northern Norway struggling with unemployment through offering mining jobs to young men and bringing them to Svalbard. Henrik remembers that at that time, the miners would know their colleagues from the mainland, people living in the same village or in the same valley. Later the shift system changed into two weeks on and two weeks off, and more people started to commute, keeping their families on the mainland and the mining job in Svalbard.

Before we continue with Henrik's story, there is a digression to follow for a little while. The issue of temporary stay was a heated topic at the time of my fieldwork as well, but by then related mostly to people who work in tourism, research, and local and state administration. The complaint that people do not truly live in Longyearbyen – that they take advantage of a financially convenient job, keep the core of their private lives elsewhere and thus do not engage with the community as they would if Svalbard were their only home – is not new to the island. During the 2000s, it was the Norwegian miners who increased the transience of the Longyearbyen population. In the following decade, it was Norwegian, but also international workers, looking for job opportunities, who further decreased the town's communitification potential (see discussion in Chapter 7).

In the 2010s, everything turned upside down for the privileged miners. First Norway invested NOK 1.2 billion in building up a new mine called Lunckefjell in Svea in 2012–13, promising 10 million tons of coal would be mined in the coming years. The mine was opened in 2014, but about a year later Norway decided to close the mines in Svea Nord for good and the company started firing people. One month after my arrival in Longyearbyen, an editorial by Hilde Kristin Røsvik in *Svalbardposten* caught my attention:

> Store Norske does what the state wants but we are sure that the clean-up job they have now started hurts. The optimistic period when Lunckefjell and Svea were built up belongs definitively to the past. The mining community in Svea will never be the same. It's over now. Almost new, unused mining material and equipment from Svea Nord and Lunckefjell will be sold. Mining companies from many countries show their interest. Material can be sold for a dumped price, end up in coal mines in countries such as Poland or Turkey, and make their mining industry even more efficient. The Norwegian authorities have decided that all the traces of the mining industry should be deleted from the surroundings of Svea. Cost what it must. The bill will go up to at least NOK 2.5 billion.[2] One justifies this through having regard for the environment

---

2.   The project turned out to be cheaper than expected. The total costs when the clean up was finished in 2023 reached NOK 1.6 billion.

– Svalbard should be a showroom of a well-protected wilderness. The central authorities do not speak aloud about the geopolitical background that most probably is at least equally important: to make it impossible for Russia and China to take over Svea and re-establish the industry. [I]n this way, we can just be done with the 'problem'. Couldn't they call things by their true names [...]? Many believed in this industrial venture but strong political powers worked against it. Retrospectively, it all seems totally meaningless. (Røsvik 2019)

The industrial optimism behind Røsvik's diatribe does not address the global problem of coal extraction that makes overheating possible, with both consumption and environmental damage accelerating worldwide in an unequal and unjust way. This is a local voice expressing discontent with decisions being taken in a non-transparent way, as if people did not matter. As Røsvik points out, there is a double bind in the embarrassing history of Svea and Lunckefjell, where billions were invested in establishing the industry only for it to be closed shortly afterwards, and even more billions invested in cleaning up the site and 'giving it back to nature'. The government justifies the decision to abandon coal mining in Svalbard through the negative environmental impact, which is an obvious standpoint to take from the outside. For example, WWF Norway (national branch of an international NGO without any local connections) delivered an assessment in 2010 to the Governor of Svalbard:

Norway must pick up the difficult long-term task of turning the big sectors (oil industry, Pension Fund investments, etc.) to a sustainable direction. At the same time, we have to point out measures that have little importance but can be done easily and quickly and will contribute to Norway's international effort to foster sustainability and reduce emissions. It is important that Norway does this because it is important that all countries do it. To sum it up, such 'easy' measures will contribute to global reduction of emissions but it is even more important that it will firmly confirm our will to actually do something. To eliminate coal mining in Svalbard is a good example of such 'easy' measures. (WWF Norway 2010: 2, translation mine)

The text continues, listing arguments for putting a halt to coal mining for environmental reasons, and proceeds to give a statement about the social impacts of the desired decision:

> Almost until 1980s, Longyearbyen was exclusively a company town, without any other industry than just coal mining, without private housing, hotels, shops, etc. Also the rest of Svalbard's settlements, Barentsburg, Pyramiden and Svea were pure company towns. Since then, Svalbard has changed significantly into a place where research, education, tourism and private business, together with management, dominate. The Norwegian population has doubled and the Norwegian presence is making itself more valid in Svalbard and internationally than ever before. (WWF Norway 2010: 3, translation mine)

What is crucial to notice here is the clash of scales embodied in the description of the measure as 'easy' and of negligible importance. For the 'little people' of Longyearbyen, the abrupt change of course was anything but simple and unimportant.

Røsvik's critique illustrates two fundamental issues: closing a newly opened, well-equipped mine in Svea while the country, with just a few millions of citizens used to an astonishingly high standard of living, is still in need of high-quality coal for metallurgical industry; selling eqquipment to countries with lower environmental and social standards; and in addition keeping the national economy running on extractive industry wholesale, is locally seen as a sign of hypocrisy. An act that globally might work as a bonus point for Norway pushing for green growth works locally as a proof of disrespect towards the people of Longyearbyen and Svea.

Second, environmental concerns might not be the most strongest argument for turning a blind eye to the wails of the miners and their families, similarly to the viscous climate change discourse explored in Chapter 3. In fact, the big geopolitical game, namely making sure that neither Russia nor China will be able to claim interests in mining (be it coal or anything else) in Svalbard if Norway sets the bar high enough, was probably decisive (Avango et al. 2023).

Time to get back to Henrik's story. 'So we turned from being in the centre of attention into something like, "we REALLY want to live without them, but right now we still need them for the coal at the power plant". You see? It's really like sloping down.' Henrik believed that the developments would proceed and that, at some point, 'the miners will be totally out of here and the university community and tourism will just keep growing. [...] One group will be totally replaced by another.' The fact that many aspects of life in the company town are already lost for good makes many people sad, 'but maybe there isn't any reason to stop it'. And then 'the great acceleration' popped up in Henrik's thinking aloud: 'But you can see it so clearly here because it is a speeded up version of what's going on in other places of the world.' He understood that coal mining is not the industry of the future, he was not a climate change denier but he voiced what I heard from many others during my fieldwork.

> *Henrik*: Now that they are closing the mines in Svea, we just had the feeling, like … ah, we could just have finished them. Okay, we're not opening new mines but maybe we could have finished the job, emptying them and closing them down while we were still working in Svea. It would have been much cheaper than this very drastic – dong – the closure that we see now. And I think that we all see it is wrong to burn off coal to get energy and to release CO_2.

He then explained what other kinds of production coal is used for, how exceptional the quality of coal from Svalbard is and why it is annoying that the debate is portrayed as black-and-white (like the miners' portraits in KB). 'Maybe there is a grey zone.' Henrik was trying to think extraction beyond extractivism, imagining a path to an alternative greener future where people performing the transition could feel some sort of ownership and participation. The path chosen from above had an alienating effect.

Henrik was also aware of the symbolic meaning of the gesture and acknowledged that Norwegians are proud of being able to manage this vast area of 'Arctic wilderness' where coal mining no longer fits in with this image.

> *Henrik*: Everything here in Svalbard is politics. It's the politicians on the mainland who are making the politics so we are just getting results of that, you can say.
>
> *Zdenka*: To what extent do you think people who live here have any kind of influence? How much can they indeed decide on what is going to happen? My feeling is that a lot is actually decided elsewhere by people who have nothing to do with Svalbard.
>
> *Henrik*: Yeah. And I think it has been like that all the time. But as I said before, people just didn't care but now they are forced to care. [laughter] We are so few people and there are so many other big interests on the mainland, putting out signals … […] It would be wrong to say that we are suffering from it, it's not true, but our lives are just a result of it. It's

very little influence that we have. But we are having a lot of visits of politicians and other groups that are coming to see.... But they are more interested in going up to see the glaciers calving and saying like 'oh, that's climate change' [embarrassed laughter].

Henrik, just like many of his fellow miners, acknowledged that climate change is a serious threat. But he disliked the fact that what, from the perspective of high-ranked politicians seems to be an easy and low-cost green gesture, changes his world profoundly.

In another conversation I had some months later, with Anne, a lady in her sixties whose life is deeply rooted in Svalbard, when we turned to the question of politics and coal mining, I told her the anecdote about a miner who got slapped in his face for doing such a dirty job. Anne reacted with determination:

> *Anne*: But we lived in the time when [coal mining] was necessary so I don't think you should have a bad conscience even if it contributed to environmental change so.... Actually, the function of this place is politics, it's only that. And in order to keep the politics going, we needed coal mining. [...] I think that we should have taken the coal out, used the money for research that can lead to new energy sources. [...] So that we can be proud of it. 'Just wait! We will turn this into something good!'

Towards the end of our conversation, Henrik mentioned that when the case was still open, the local debate about the mines was heated. But once the final decision came from the mainland, the topic disappeared from the agora and the focus shifted to tourism and its negative impacts on life in town. When asked about his personal opinion on tourism, he commented: 'I think it's the necessary evil.' Then he brought up the issue of placing blame: 'Maybe people are thinking about us the same, basically? Maybe I am doing exactly the same thing to other people as what they have been doing to me.'

Ethnography tints the 'coal mining problem' (Saville 2022), painted black-and-white from the outside, with colour. Coal in Svalbard cannot be understood only through the environmental perspective. It was a political tool that no longer served the cause and so it was eliminated. The prices were too low, the economic loss too high and there was no point in keeping a 'dirty' industry running in a 'clean' place like Svalbard, as this would be harmful for Norway's global PR. The way in which the decision was taken and put into practice deprived those dependent on the industry of any form of agency, and made it difficult for people to narrate a meaningful story of the place they inhabit. During lunch at *Huset* ('The House' – a cultural centre) in the middle of the Optimal Tourism Balance workshop held in Longyearbyen in September 2019, I chatted with a Visit Svalbard employee. He was concerned: 'It's not interesting to come to a place that only has one story to tell – tourism. You need something real.'

A House under Every Bush and a Hole in the Jumper: Scaling Tourism

Coal mining, tourism and research and education have many things in common on Svalbard. The most obvious one is that all these activities are primarily meant as political tools to achieve a certain goal – to confirm and foster the Norwegian presence and sovereignty over the strategic area. But there is one feature that makes coal mining and tourism profoundly different: While it is impossible to run a family coal mine on Svalbard, it is perfectly possible to run a family tour company.

In order to understand the elevated level of dependency that I encountered during my fieldwork, we need a brief introduction to the history of tourism on Svalbard, paying attention to the role that 'big powers' played in it. From a certain point of view, the history of tourism means the history of the human presence on the archipelago. 'Mister Longyear was here as a tourist,' I heard many comment on the issue. There is a joke circulating: 'Who is local in Longyearbyen?' 'A person who arrived one plane earlier than you.'

Depending on how we define tourism, it might be possible to identify features characterising tourism already in the ventures of the sixteenth, seventeenth and eighteenth centuries around Svalbard. In the context of overheating, though, it is more fruitful to look at tourism in the light of the changes that took place in the nineteenth century, the era of industrialisation, of wider and quicker travel possibilities and, last but not least, leisure time. On the website of Hurtigruten Svalbard, the visitor can read the following summary:

Hurtigruten Svalbard is Svalbard oldest tour operator. From a travel agent which helped miners to head south, we are now a full service provider here to help you travel north. The first tourists who came to Svalbard were rich Europeans who came to hunt, and often wrote books about their experiences. [In 1896], [...] the company set up a prefabricated hotel in Longyearbyen. The building had accommodations for 30 guests, and was the northernmost hotel in the world. [...] Two years later the tourist route was put on pause, and the hotel was sold to the new local miners, Arctic Coal Company. The building was used as a shop, until it was burnt down by the Germans in 1943. Even if the hotel did not exist for long, it was an important part of Norwegian industry on Svalbard, before the mining industry really took off.

There are two noteworthy things about the representation of tourism by this large player on the local market (in fact much larger by far than any other). First of all, it portrays itself as a local operator with a historical bond to the place. At the time of my fieldwork, it was widely known that Hurtigruten's CEOs sit in London and might be tempted to sell out to the Chinese (which did not happen; instead, the whole housing portfolio of Hurtigruten Svalbard was sold to Store Norske in 2021, reconfirming the company's role in town). Second, it presents tourism as a historically more traditional activity than coal mining. In the previous chapter, I elaborated on 'taking something out of the place' and how, in such a narrative, tourism should not be understood as opposing coal mining or replacing it, but rather as a complementary though softer extractive industry.

In the governmental document issued in 1974–5 discussing the status quo and nationally desired future development of the area in the geopolitical context of the Cold War and search for oil and gas, formulations about tourism were more than hesitant:

> It will however hardly be in Norway's interest to turn Svalbard into a distinctive tourist land, both because protection plans might be endangered and because the economic advantages of tourist traffic in Svalbard are considered as modest. [...] Svalbard's untouched and vulnerable nature and the significant protection interests related to the whole area, not only national parks and nature reserves indicate that we have to avoid substantial commercial tourism using motorised transportation means both in terms of land, water and air. (Norwegian Ministry of Justice and Public Security 1974–1975: 34, translation mine)

An important milestone for the years to come was opening the airport in 1975. A slight change of course in the national strategy can be detected 10 years later, in the next governmental White Paper discussing Svalbard.

Calculating the scope of tourist traffic in Longyearbyen showed that it will be difficult to achieve business profitability. This would only be possible through enabling tourist traffic on a larger scale, which has not been considered as desirable. The committee therefore saw no other possibilities than supporting the plans financially from public sources. This meant a grant of 22 mil. Norwegian kroner from the state budget in 1978. (Norwegian Ministry of Justice 1986: 33, translation mine)

One more thing happened in that period. In 1976, a girl of Sámi ancestry, then in her twenties, moved to Svalbard from Northern Norway, just to experience it for a year or so.

I am cold on my feet, my cheeks still red from the nippy wind that blows in Adventdalen and carries away the few snowflakes we received with joy that April morning, revealing the raw face of the stony tundra where hardly any skier or dogsled can go but where snow scooters still buzz back and forth on the brownish path of a thin snow cover. I switch on my audio-recorder, stutter a few sentences about my research but my partner in conversation today has no time to wait for my questions. She listens to my simplified explanation of overheating with maybe a bit of impatience, and I notice she is wearing a pair of boots and a woollen jumper that certainly have not been purchased recently. I ask Berit Våtvik about her motivation to move to Svalbard.

> *Berit*: The youth nowadays, they are globetrotters but we couldn't travel like that. [...] It was the kids from rich families who could afford travelling. [...] So it was my adventure, to do something else before life begins. And polar history! I was born in Northern Norway so these were the stories of my childhood. I was interested in outdoor life and curious to see where it started.

Berit worked for SNSK for many years, both in Svea and Longyearbyen, found a life partner and had three children. It was only in the 1990s that the volume of tourism increased significantly and Berit thought it might be wise to create a job on her own. The mining company took over new communal functions and one of them was management of a small tour operator targeting locals first, helping them organise trips to the mainland. But the stream of visitors gradually grew bigger the other way too, from elsewhere to Svalbard. Earlier, it was mostly people visiting those who already lived and worked in Longyearbyen but when the first accommodation facilities mushroomed, encouraged financially by the state, guests from all around the world started showing their interest in the unique place repeatedly described as 'untouched and vulnerable'. Berit was not the only woman who decided to kick off her private business in the predominantly male society. The company was first called *Svalbard Miljøturer* (Svalbard Environment Trips) and was based on offering trips on dog sleds – something that Berit and her family were fond of. It felt natural to combine one's passion with a small-scale employment possibility.

There was an important value embedded in the foundations of the humble family business: silence in nature. 'It's a value that cannot be translated into money ... you can say that you get growth in terms of *Homo oecologicus*, not *Homo oeconomicus.*' I sense that these terms, *det økologiske menneske* and *det økonomiske menneske*, play a crucial role in Berit's life philosophy. 'We were educated by nature.' Work with dogs requires your effort 24/7 so there was no point in spending time in town – the family business was part of a holistic lifestyle. 'The world came to us,' says Berit, referring to the guests they were meeting on trips. But she also mentions that they were challenged by people pointing fingers at them for using a car, a computer, all in all – at them being part of the world that was changing and speeding up. Being 'put up against the wall', the Våtviks felt it was wiser to change the name of the company – *miljø* (environment) should not appear in it anymore. *Svalbard Villmarkssenter* (Svalbard Wilderness Centre) came into being.

> Berit: And at those times we also talked about this thing with coal. [...] But nobody said like, metallurgical industry versus burning it up and warming up our houses, right? It was just coal, right? [...] But we spoke about using coal to produce important things. [...] Should we just burn it up if it took the planet 60 million years to create it? It's like, you empty a bank. But today, it's just economic growth so everything's lost. People don't reflect on that even if they destroy the whole world.

The computer ventilator buzzes every now and then and I feel the warm air coming out. Berit is playing with her fingers and keeps adjusting her colourful hat.

> Berit: The authorities in Svalbard, they are great people, diligent and everything, but there is no long-term plan. They shot themselves in the foot ...

Zdenka: What do you mean by that?

Berit: We haven't followed up as we should have. There is only growth....
Now they're trying, instead of demanding something at an earlier point
in time, to close the door, here on land. But they flood the fjords!

She is referring to growth not only in land-based tourism but also on the
sea. In 2019, land-based tourism attracted about 77,000 visitors to the
archipelago, with turnover of NOK 851 million. Conventional cruises
brought about 40,000 passengers but had a turnover of a mere NOK 31.8
million, while expedition cruises accommodated 22,000 tourists with a
solid turnover of NOK 97.4 million (Visit Svalbard 2022a). The Covid-19
pandemic brought a rupture in tourism developments but in 2022 tourism
seemed to be back on track. The future of cruise tourism in and around
Svalbard is uncertain because of strict environmental regulations soon to
come into force but the potential for a boom is indisputable.

Berit had a lot to say about motorised transportation for leisure. At the
time of my fieldwork, I often had the impression that snowmobiles were
crawling around the town like giant beetles, noisy and smelly. They were
everywhere, parked randomly, fancy new ones and shabby abandoned
ones, and there was a whole army of them on the sea shore, owned by
the two biggest actors, Hurtigruten Svalbard and Svalbard Adventures.
Berit told me about the process of drafting and sending input for the
Svalbardmiljøloven (Svalbard Environmental Law) that should have guar-
anteed vast areas would be protected from motorised transportation. The
point was to make sure that companies offering non-motorised trips in
the Arctic terrain (on foot, on skis or with a dogsled) can be competitive. But another, equally important, goal of the measure was to protect
those areas for local people, for their private recreation, on the condition
that you are willing to make a little effort – reach a glacier that is available to everybody without a scooter. It took a long time to prepare the law.
But then some people weren't satisfied; there were some petitions and
eventually the law was changed, saying that motorised means of transportation are allowed during the winter season until 1 March. 'It was the
authorities who destroyed the plan. None of the people still live here. We
are very few who know the history.' The aim was to 'protect nature but
also protect local jobs, local jobs, you know', Berit's hand hits the table.
'A bit of empathy, not only for each other – "I wanna drive scooter!", "I
wanna drive dogsled!" – but also for those who come after us.... There

was a long-term plan that got destroyed within a few days.' Her trust in local authorities was shattered, if not lost altogether. There was a local initiative of people with roots in Svalbard who had a long-term vision and a value system to base it upon, but their voice was too quiet compared to those of bigger and more powerful stakeholders; those who profit from tourism on a larger scale than a family company run by a handful of people with a simple lifestyle, who care about the dignity and happiness of their dogs, whom they do not treat as working tools but as life companions.

Before I leave, Berit says – without me asking – that her mother knitted the jumper she is wearing. Her fingertip peeps through a small hole. She is not good at mending them but a friend promised to do it for her.

Just two days after the visit I paid to Berit, I went to *Huset* to watch *Révyen*, an amateur revue summarising the year that just passed with irony and satire; a string of sketches written and performed by locals, with lots of music, singing and jokes for insiders. I later asked the authors to share the lyrics with me and I couldn't stop thinking about the hole in Berit's jumper and her worry about local jobs when reading the screenplay to the sketch entitled 'House under every bush' (*Hus under hver busk*, written by Jovna Zakharias Dunfjell and Frida Vestnes). A tired tourist from Stavanger wanders around Longyearbyen and always ends up in a facility owned by Hurtigruten Svalbard: Camp Barentz (wilderness experience centre), Green Dog (dog yard), IGP (scooter rental facility), Radisson Blue Polar Hotel Spitsbergen (accommodation), 78° Longyear (sportswear shopping centre), Coalminers Cabins (accommodation), and finally the most expensive hotel in town, Funken Lodge, where the worn-out pilgrim is asked to pay NOK 3,000 for one night and exclaims: 'That's a lot! Are there no alternatives in town?' The Hurtigruten Svalbard employee answers: 'Unfortunately there aren't! In any case I haven't heard of any. There is just Hurtigruten…. Welcome to us!'

Who Can Pull the Emergency Brake

There is Norwegian coffee in a thermos on the table, too 'long' for my taste but after one year in Longyearbyen I am getting used to it. Annicke, one of the Visit Svalbard employees, is explaining to me how she perceives the current situation regarding developments in tourism. By the end of 2019/ beginning of 2020, several signals were sent out to Svalbard from Oslo that could potentially change the business. First of all, a ban on heavy oil in ships cruising to Svalbard. Everybody applauds; there is no disagreement about this one. Then, certification of guides. Unqualified guides and irresponsible companies were a heavily discussed topic at the Optimal Tourism Balance workshop some months earlier, and it was obvious that something needs to be done about people and companies coming to Svalbard without any local anchorage, profiting from the booming industry without making any contribution whatsoever to the place bearing the burden of increased traffic. There might be a catch in the issue of guide certification (who decides, who evaluates, who issues the certification, who pays whom and how much, etc.) but this measure has also been welcome as beneficial, with a slightly bitter feeling that something like that should have existed some 10–15 years ago already, before the archipelago was flooded by all those astute polar cowboys who just 'lick off the cream' (*skummer fløten*). Finally, housing available and owned by state-funded entities such as LL or UNIS is supposed to be formally sold to Store Norske or Statsbygg.[3] The reality is more of the state buying its own property and letting the same state manage it centrally. During the Norwegian Prime Minister Erna Solberg's visit to Longyearbyen in February 2020, she gave an explanation going from vulnerability to activity and back: 'We have to take care of this vulnerable environment and that means limitations in terms of what kind of activity you can perform here. We also have to make sure there is not too much activity in total as the environment is vulnerable' (Wiersen 2020, my translation).

Annicke says to her it seems they have opened the dam without thinking it through, tourism exploded and now they are 'trying to pull the emergency brake and the train is creaking and screeching'. In the White Paper delivered in 1991 (Norwegian Ministry of Commerce 1991), tourism was presented as a well-established and growing industry in Svalbard, and described as desirable but in need of regulation. Many things that I heard people complaining about (the tricky question of quantity, missing infrastructure, nature and cultural heritage exposed to risk, etc.) are already identified in this 30-year-old document. All the issues that started to be visible in the 1990s, at that time met with modest interest on the part of the locals and slight worry on the part of the central authorities, accelerated astonishingly during the first two decades of the new millennium. The result is a local economy dependent on global mobility, foreign capital that is hard to track, and exploited international workers.

---

3. These transactions were completed in 2021/2.

Shortly after my arrival to Longyearbyen, I became puzzled that the supposedly thorough long-term governmental strategy was locally perceived as unconvincing. The pandemic starting in 2020 shattered tourism worldwide and, in the Norwegian context, hit Svalbard hardest. In 2021, the destination management organisation Visit Svalbard conducted an online survey among Longyearbyen residents about their perceptions of tourism (Visit Svalbard 2022b). Of the respondents, 74 per cent were positive or quite positive about tourism (compared to 90 per cent in 2015). The positive attitude prevailed among those who had lived in town for less than six years, comprising more than half of the respondents and about two-thirds of the population. The vast majority (74 per cent) were negative towards cruise tourism specifically. Among the positive impacts, people indicated employment opportunities, the urban and cosmopolitan vibe of the town, diversification of services and leisure activities, and regular and affordable flights. The credibility of Longyearbyen's environmental profile as a tourist destination turned out to be blurred, too; only 13 per cent believed Longyearbyen was a sustainable destination. The survey showed residents' awareness of the industry's environmental paradoxes, overtourism related to cruises, class divides, rising inequalities in working and living conditions, and insufficient local participation in shaping Svalbard's tourism future.

During my conversation with Jens, a senior scientist working at UNIS, we touched upon the discrepancy of opening up for global tourism and being taken by surprise of its supposedly unforeseen consequences.

Jens: It's correct to say that tourism has been pointed out as one solution. That is very common for a lot of rural parts of Norway. [...] But what they don't do is put a specific meaning into, like, what do they mean by 'sustainable', what do they mean by 'local' and so ... I think a lot of the things written in the White Paper are generic. It is nothing special for this place. And then, looking at it from the perspective of the tourist industry, they mostly look at the potential. So if people are willing to pay for it, they will build it up. If people are not willing to pay for it, they will not build it up. For them it's simple.

Jens pragmatically stated that tourism is nothing new to Svalbard: rich people were eager to spend money on visiting the archipelago in the past just as they are in the present.

Jens: So they wrote it in the White Paper and they have been saying it for many years. And now the tourist industry has been able to build up [...] and we start to see the problem. Because it's been developing too fast, it's too many, they are using the infrastructure that was meant for the local community, you know, all these issues. So in a way, the authorities have got what they asked for. And it might be difficult for them to say afterward 'yeah, we didn't really mean it'.

In Longyearbyen, 'little people' are influenced by 'big powers' that seem both extraordinarily near (local politicians often brag about having a direct link to ministers and national party leaders, which politicians from an equally small community on the mainland could only dream of) and endlessly far away (those in charge sit three hours away by plane and have other than local interests in mind when taking decisions). When I spoke to people working for LL, be it clerks employed in the administration or local politicians, my impression was that the limited political power was accepted as a fact, and not something that could or should work as an impetus for political mobilisation (for more on this issue see Chapter 7). People would express modest concerns about the turnover among *Polaravdeling* members (a body consisting of representatives from the ministries that share responsibility for Svalbard), and also a wish for a clearer long-term strategy that would not combine contradictory goals and analyse possible consequences before the impacts of the measures start hitting too hard and too fast. The question of agency – playing an active role in what some call 'only a geopolitical theatre', where people serve as 'the government's puppets' – became central in my endeavours. In the next chapter, we shall stay with the trouble of tourism and zoom in to those most involved in the business through their physical labour, living the paradox of Svalbard as their daily bread – the tour guides.

6

Sustainability with a Footnote: Leaving Out Justice

During an interview, a non-Norwegian tour guide used the term 'little people' to express the irrelevance of local mishaps in the light of the big/ high political game Norway is playing when exercising sovereignty over Svalbard. I was interested in the lifeworlds of the tour guides for several reasons. Their grouping, in emic terms self-labelled as 'the guiding community', came into being in Longyearbyen thanks to the fast developing tourism, which could not do without them. It is difficult to estimate how many and who they are; while there are many incentives to register with the local Norwegian Tax Authority office in Longyearbyen, which is basically the only formal criterion for turning you into a 'local', no stick is used in practice to counterbalance the carrot if people forget or do not bother to unregister when they move away from the archipelago. Primary employees include destination managers, tour company administrators and guides. Secondary employees are those in related services, such as hotel managers, bartenders, cooks, cleaners, shop assistants, etc. There is a racialised job market: for example, in the cleaning sector a majority of the workers are from Thailand and the Philippines (see Chapter 8), while it is almost exclusively Norwegian nationals who work in tourism management.

It is not rocket science to figure out that tourism labour is different from the mining labour dominant until recently. An indisputable correlation exists between the tourism boom and the rise in the non-Norwegian population, yet this is hardly ever mentioned in strategic documents about tourism. People working in bars, restaurants, hotels, and most tour guides, migrated to Svalbard from outside Norway.

According to a survey by AECO (2022), of 522 tour guides operating in Svalbard, over 90 per cent come from outside Norway. Only 62 of the 522 are Svalbard residents. The reason for such a low number is that tourism on Svalbard is heavily dependent on the seasons, which also determine the type of guiding performed. For example, tour guides who work for the cruise industry can only be around in the summer season; according to the leader of the Svalbard Guide Association (personal communication), about 300 tour guides who are members of the association are employed in the Antarctic when the polar night turns Svalbard into a dark nest. People who work in sectors other than tourism thus often perceive tour guides and tourism workers generally as transient.

From the outside, tour guides might look like tourism's replaceable workforce, miners of polar experiences and memories slightly outnumbering coal miners in the extractivist town of Longyearbyen. From within, they constitute a diverse assemblage of both globetrotters and home seekers (Sokolíčková and Soukupová 2021), who are rather young, often highly educated (Bachelor and Master's degrees are common, and PhDs not too exceptional), locally engaged and, at the same time, international and cosmopolitan. In stark contrast to the miners, who are remembered as those who 'built the town of Longyearbyen' in the predominantly Norwegian past of the settlement, the tour guides' role seems more ephemeral, less acknowledged as the town's economic cornerstone. In fact, their accounts of what life in Longyearbyen is like for them made me wonder whether the key agents of mediation (La Cour 2022) of what 'the Arctic' can mean for those who come to visit are at best tolerated and at worst ignored.

My curiosity about the lived experience of the tour guides had a personal component; when we moved to Svalbard and placed our youngest son in the kindergarten after the initial five months when he was at home with my husband, Jakub started to look for a job. After a few unsuccessful efforts to connect to the UNIS environment, which would have been most natural for him given his educational background and professional expertise, he landed in the guiding community, working occasionally for four local tour operators, but mostly for one of medium size.

Our family life was impacted by the dilemmas a tour guide must navigate only for a short period of time. It took just a few weeks from when we first started to talk about Covid-19 to the point when all the tourists were evacuated from the island and the whole industry was put on hold, resulting in Jakub being laid off, losing his income until the end of our stay; the first busy season 'back to normal' was spring 2022. Yet even the few months of partaking in the tour guides' world were challenging.

With stars in my eyes and a suitcase in my hand
Looking forward to a new start at life in a new land.
[...]

Half of my salary was spent on rent;
I was seriously considering living in a tent.
[...]
You probably know how this story will end,
I gave up my job and life here, I just had to disappear.

We did not have to consider living in a tent as I was allowed to rent an apartment through UNIS. Camping would have not been an option for a family of five, but we knew people for whom the camping site became the only option when they could not afford any other housing. Among tour guides, periods of couch surfing at friends' places was a common strategy. In 2018, the guiding community founded their own organisation, the Svalbard Guide Association, to fight against long shifts, low salaries, physical and psychological fatigue, and unpaid overtime. A public Facebook page, 'Homeless 78° North', was created in 2019 and has over 300 members, including tour guides and other tourism workers. The page displays a poem, from which I share a few lines above, written on an image of Longyearbyen's photogenic *spisshusene*, colourful houses with pointy roofs often featured on postcards tourists can buy in Longyearbyen. 'You have a picture postcard, perfect, it's being sold as what the town is – but what we are is something quite different,' a participant of a focus group commented in June 2020 on the town's branding.

Precarious Lives: Keeping the Wheels of Tourism Turning from the Margins

The environmental paradox of Svalbard tourism, and how people involved in it negotiate meanings and values bearing in mind what is happening in environments worldwide, is self-evident and also documented in existing literature (e.g. Andersen 2022; Saville 2022). Tourism significantly increases air and sea traffic, which causes higher carbon emissions and leads to ocean pollution. Higher volumes require higher standards in infrastructure, which again pushes up energy consumption, predominantly secured through a local coal power plant. The most popular means of transportation – boat, snowmobile, and dogsled – all have environmental footprints (a comparative study from Svalbard is missing). More subtle impacts should not be forgotten: tundra is vulnerable to damage (by snowmobiling mostly, but also hiking); increased human activity can potentially disturb wildlife; residents report noise pollution by snowmobiles; intensely used landing sites bear traces of wear and tear, and so on. Environmental doubts and reflections over one's contribution to negative environmental impact locally and climate change globally through partaking in extractivist activities came up in my interviews with tour guides but also with tourists, as in the three examples cited here:

> I think it is a lot of contradictions. You have this beautiful Arctic 'untouched nature' and then the whole infrastructure and everything is based on coal mining in the Arctic. Everything has two sides up here. I come to work in this very beautiful environment but at the same time, you know, I affect it. [...] So it's very mixed feelings actually.

> And tourism, how it has developed, well, I don't know whether it is good or bad, there are always two sides. I am happy that there is more tourism because I get a job. [...] I love to be here so I am happy for that, but on the other hand, I mean, every person coming up here is coming by airplane, using mostly ships or snowmobiles in the winter, whatever, so it's not very ecological.

> Well, in fact our consideration was whether we would come at all, whether we are contributing in some way to climate change by coming and visiting. So I was concerned that we might be doing damage.

These mixed feelings about Svalbard, both represented as a climate change victim and pioneering sustainable solutions hub while itself being fundamentally unsustainable, mostly touch upon the environmental layer of the paradox. It has already been appropriated by the official discourse, depoliticising tourism:

> Some may well ask: How can Svalbard be a sustainable destination when its energy comes from coal mining, most consumer goods are imported to the island, there are daily flights to and from the mainland and in the summertime large cruise ships arrive with thousands of guests on board? We often hear this as arguments for why we are not a sustainable destination. However, this is what we must deal with in the current situation. Despite the environmental paradox in Svalbard, we are working continuously to improve the conditions in this setting. [...] We implement local measures, make wise environmental choices and, not least, increase awareness of sustainability among the tourism industry, the local population and our guests. (Visit Svalbard 2022a)

Rob Nixon's (2011) appeal to reconsider how we understand violence, drawing on Galtung's concept of structural violence but bringing in forms of pressure that occur 'gradually and out of sight [...] neither spectacular nor instantaneous' (p. 2), resonates in the debate about environmental justice. Might it be fruitful to invite the analytical tool of slow violence to a conversation about tourism, too? Nixon applies it convincingly when discussing tourism in Africa, unable to disentangle itself from colonial imaginaries looking for authenticity and wilderness while local populations are systematically disempowered, displaced and essentialised. The case of Svalbard is different but not unrelated. Svalbard's environments are drawn into the performative act of tourism, in which 'nature' is represented as spectacular, both threatening and vulnerable, authentically Arctic and pristine. The tour guides lived the acceleration of tourism interrupted by the pandemic in 2020 as a succession of struggles they felt were somewhat unspectacular and ignored, as a grouping that resembles Nixon's 'disposable people'.

> *Toni*: I think this place only loses if it continues to be Norwegian. There are two things so precious about this place: the nature, and being international. I find it absurd when the Governor's Office gives a lecture for tour guides on environmental protection, and it is in Norwegian. But the only reason why I didn't stand up and say 'fuck you all' is that, unfortunately, this place does not allow for that; here you have to be careful about what you do.

Toni was one of the most skilled tour guides in town. With many years of experience and the highest possible qualifications in a wide range of guiding activities, she felt confident about her position in the company. Toni belonged to those who return for particular seasons instead of staying all year round. She worked full-time and had a varied schedule, going hiking, kayaking, leading one-day but also advanced multi-day expeditions, both on skis and using snowmobiles. She saw herself as part of the guiding community, where she was respected as a knowledgeable local others would call to ask about conditions in the terrain and advice on how to plan trips. Activities outside town require expertise and the ability to take responsibility for safety when it comes to polar bears, sea ice, a driving or hiking route, avalanches, or glacier crossings. Toni loved what others call extreme sports, but for her that was the norm – she worked as a tour guide because experiencing nature through physical activity was her life anyway, so to make a living like that was the only thing that made sense.

Though Christian (2017) studies tourism labour in Asia and Africa, the situation resembles Svalbard: 'There are just a handful of permanent employees who receive reasonable salaries with health and other benefits' (2017: 815). It is locally widely known that tour guides have suffered from a prolonged and politically ignored housing crisis. They earn on average NOK 160–230 (€16–23) per hour; their efforts to organise and stand up for their rights are insufficiently supported; and many cases were reported where a guide pointing out serious issues (discrimination, safety issues or work exploitation) would be threatened by the employer with losing their job. This population segment is critically important for keeping the wheels of tourism turning, but it challenges Norway's representation of Longyearbyen as a stable Norwegian community consisting of affluent families (Norwegian Ministry of Justice and Public Security 2015–2016; Pedersen 2017).

> *Ole*: Here you have to pick your fights because there are too many. An example from last year: as a guide you sign three work contracts a year, one for the summer, one for the winter, and if you are very lucky you get one also for the autumn. So in the contract for last summer I had a full-time position, 162.5 hours a month. [...] It should have lasted from 15 May until 15 October. On 1 July I was missing about 30, 40 hours. I asked the boss and he was like, 'Just chill, it's 162.5 hours a month on average, you will catch up within the six months.' [...] By the end of August it was about 65 hours I was missing, and he kept telling me, 'Calm down, it will settle.' [...] When we arrived at 15 October, I was missing over 120 hours, that's almost a month's salary. I came to the boss and said, 'Here is my contract, I was promised a full-time job, it's in the contract, it's in the law.' And this is what I got as an answer: 'It is correct what you say. Take your contract to the accounting office and ask them to pay you for the hours you haven't worked.... But by the way, where is it you were planning to work next year?' You see? Three times a year they can fire you, and there are so many standing in the queue wanting your job. It's no problem to substitute you with somebody else. So if you don't tread on eggshells and smile nicely, they will just get rid of you.

Ole was one of the few Norwegian guides I spoke to, and he was so upset about how poor the rights of the guides were that he did all he could to move to the safer public sector on the Longyearbyen job market, and succeeded about a year after our interview.

While structural violence is silent, invisible and static, Nixon (2011: 11) seeks 'to foreground questions of time, movement, and change, however gradual [...] to engage directly with our contemporary politics of speed'. A glacier melting at an eye-catching rate will attract much more attention than the deteriorating situation of tour guides, while both phenomena relate to the business's sustainability. The guides tell stories about melting glaciers, but their experience with exploitation and political marginalisation cannot make it into their storytelling reservoir.

On the voice recording of the interview with Toni, I hear the breath of a retired sledge dog adopted by my interlocutor's flatmate. It is a common pattern among tour guides in Longyearbyen to take care of dogs too old to stay in the business of dogsledding. As a sign of respect for the past labour and meaningful life stories that deserve continuation beyond the horizon of work, people take dogs from the kennels and make sure they live through a dignified retirement, provided with food, housing and love. When my partner lost his guiding job due to the pandemic and struggled to claim his rights just like those working in the industry for many years, he felt tourism workers were less valued than sledge dogs. 'Those are taken care of.' During the interview conducted with Toni in a rather badly maintained house with small rooms rented to tour guides, Toni mentioned that one of her flatmates might need to sleep in a tent from next month on as no more housing would be available for them.

The paradoxes of tourism are far too obvious in a place like Svalbard, and to come to terms with the dilemmas, the flag of sustainability is being hoisted high. The destination management organisation Visit Svalbard published a new Masterplan in June 2022 as a framework for tourism development towards 2030, and sustainability of the business is its key theme. Thinking with my participants from the sphere of tourism (19 tour guides and 34 other tourism workers from managers, tour company owners, through marketers and sales assistants, to people working in gastronomy and cleaning), my attention was caught by an absence emerging from the data: while tourism in Svalbard has something important to say about how sustainability relates to justice, the topic of justice does not appear in the public discourse (with the exception of some good journalistic coverage) including in strategic documents, but also largely in theorising tourism in Svalbard.

Obviously, the meanings of sustainability are numerous and context-dependent (Butler 1999; Saarinen 2006), which is also a source of frustration for tourism scholars (Saarinen 2014). Since the 1970s, researchers have turned their attention to social and environmental issues within tourism (Buckley 2012). Fifty years ago, Young pointed out that a growing demand implies the need to adjust tourism to national, regional and local dimensions to avoid negative social, economic and environmental local impacts. However, it was not until the 1990s that scholarly debate around sustainable tourism started. In her review of the development of sustainability indicators for Arctic tourism, Ólafsdóttir admits that 'an agreement on what precisely is sustainable is still far out of reach' (2021: 2). Pasgaard et al. acknowledge that 'lack of conceptual clarity is also reflected in tourism practices that are (self-)labelled as sustainable' (2021: 17). Today, sustainable tourism is a heavily studied topic, fanning out over many different areas of interest. The term has become a buzzword, allowing 'many tourism developments to take place under its rubric that are less than sustainable and have been located in sensitive areas that have not all been suitable for such development' (Butler 2018: 1). Butler believes 'much greater scrutiny of what passes as sustainable tourism needs to be undertaken before it is automatically given carte blanche to have free passage to any destination by virtue of its label' (2018: 1), also encouraging scholars to stop ignoring the political aspect of tourism. To what extent is that true about scholarship focusing on Svalbard?

Literature on Svalbard tourism has a tradition starting in the 1990s, when the industry received a significant boost. Most researched is the interface of environmental protection and tourism, often in the light of complicated governance regime (e.g. Kaltenborn and Emmelin 1993; Viken 1998, 2006, 2011; Viken and Jørgensen 1998; Viken et al. 2014; Nyseth and Viken 2016; Hovelsrud et al. 2020; Kaltenborn et al. 2020; Viken 2020a, 2020b). In an article now 30 years old, Kaltenborn and Emmelin (1993: 49) ask whether 'the present tourism phase in Svalbard's history will be the last chapter in a story of limitless exploitation of the resources' – a rhetorical question followed up on by Saville (2022). For 20 years, the sustainability of Svalbard tourism has been discussed as a problem of how to 'reconcile the goals of environmental management with those of economic development' (Kaltenborn 2000: 29), balancing economic growth and environmental conservation (Hovelsrud et al. 2021). Viken addresses the environmental paradox embedded in the 'big business' of tourism (2020a: 16) and how 'those who are governed wish to take part in the governing processes' (2020b: 301). Methodologically, most of these studies are based on qualitative interviews with tourism stakeholders – thus people in 'key positions' (Viken 1998: 86) and not tourism workers. Through exploring the lifeworlds of tour guides, hoping to understand better the multi-layered change brought about by booming tourism, I started to see that the local discussion about sustainability has a footnote written in a very small font: 'The issue of justice is not taken into consideration.'

Since the mid-2000s, critical tourism studies has witnessed a move described as the 'ethical' or 'justice turn' (e.g. Blackstock et al. 2008; Grimwood 2015; Guía 2021; Rastegar et al. 2021).

According to Jamal and Camargo (2014: 11):

> justice is a challenge that sustainability scholars can ill afford to ignore. Concepts of justice, fairness and equity were a fundamental aim of early sustainability pioneers (World Commission on Environment and Development ([WCED], 1987), but in proliferating principles and developing new management techniques, we appear to have lost sight of issues of justice in sustainable development generally, and specifically in sustainable tourism.

The fuzziness of the terminology and the risk of leaving out justice issues related to tourism workers is noted by Christian (2017: 805–6):

> [T]our operators use language about sustainability [...] for brand product differentiation and marketing. [...] Standards are limited and do not include measures to protect against precarious employment, gendered divisions of labor, and emotional labor demands. [...] Codes of conduct and different labels and certification schemes promote a broad grouping of sustainable criteria surrounding environmental, economic, and social impacts and represent the dominant method for firms to signify their commitment to sustainability. [...] Yet the definition of what is sustainable is a contested field. More specifically, how, if at all, these mechanisms protect and guarantee the rights and protections for tourism workers remains an open question.

In their volume on 'socializing tourism,' Higgins-Desbiolles et al. (2022) warn that 'to think critically on tourism and to challenge its injustices may be interpreted to be anti-industry or even waging a "war on tourism"'.

Such has never been my intention. Yet I resonate with the authors' claim that 'the place where tourism occurs is not a tourism destination; it is the local community's home, their standing place, and a place of uncompromisable value' (Higgins-Desbiolles et al. 2022: 14). When it comes to what tourism should secure for tourism workers, they name 'stable work, a living wage, appropriate hours, social benefits; and enabling rights [such as] lack of discrimination, voice, and collective bargaining capabilities'. Yet, 'global tourism organized at the massive scale we have known until now is impossible without the availability of both a numerically substantial and functionally flexible workforce of migrant workers' (Salazar 2020: 4); that is, disposable people who are not included in the discussion about what sustainability means. As Salazar (2020) shows further, every place needs a self-tailored and locally anchored strategy. Ólafsdóttir (2021) argues for the importance of public participation in the debate where local and expert knowledge should be equal. Higgins-Desbiolles et al. (2022) urge central governments to recognise the authority of local community over tourism. Lindberg (2020) proposes participatory tourism (*deltagende turisme*), calling for more research unpacking what that could mean in the Norwegian context.

In their commentary on how to create a justice framework for a post-pandemic tourism recovery, Rastegar et al. acknowledge that crises exacerbate existing inequalities and '"tourism-dependent communities" have now become "communities in crisis"' (2021: 1). Bringing justice to the centre of global change for tourism futures means we cannot go back to the unjust and exploitative structures of tourism. As Rastegar et al. put it, 'recovery strategies cannot be placed in an ethical vacuum' (2021: 2). The justice turn includes both addressing the socioeconomic impacts on voiceless and marginalised groups, as well as the ecological impacts on diverse ecologies: in short, a shift away from commodification and depoliticisation in tourism. However, 'research related to justice and tourism is in its infancy and much awaits research and praxis to build a robust knowledge base and weave tighter just tourism futures' (Jamal and Higham 2021: 154).

In Longyearbyen, it will be difficult to achieve a justice turn in the essentially extractivist business of tourism without tackling the rights of tourism workers, including the tour guides. In her unpublished Bachelor thesis, Hokholt Bjelland, who herself is educated as a guide on Svalbard and has worked in the industry, documents the low social status tour guides complain about:

It is actually not acknowledged in any way apart from some big words such as […] 'yes, the guide is important,' and 'the guide is the key to everything,' but this means nothing and it is only empty words that one says to make the guide feel visible. It's very disappointing. (quote cited in Hokholt Bjelland 2019: 25–6)

In the new Masterplan for sustainable destination development, tour guides are mentioned twice; once when stating that focus on relationship between the tour guide and the guest should increase, and once when listing which services should be of higher quality, such as 'capacity and quality of equipment, guides, and experience selling points' (Visit Svalbard 2022b: 35). During a focus group held in June 2020 as part of a joint project of the Svalbard Social Science Initiative with the local architecture studio LPO Arkitekter, we invited guides to comment on how they perceive their situation. Their social and economic situation, and the feeling of disempowerment and being underprivileged was pressing:

I don't entirely feel very welcome, being non-Norwegian and also being a guide. Most guides here are seasonal, so I don't think our opinions matter so much, because we're just coming in and out, we're very transient. We're here when the work is here, when the opportunity is here.

When asked about the town's identity, another guide elaborated:

If I were to describe it to my guests, I would say it's a small town, where […] people in general focus a lot on the environment and outdoor life. And it is one of the most peaceful places on Earth. But if I would describe it to someone else, I would say that it is more like a Norwegian colony. It's controlled by people who have maybe never even been on Svalbard, they don't know anything about the Arctic environment. […] Seasonal workers cannot vote, they cannot say anything, they cannot make any difference, even though they have been working here for the last ten years.

In another unpublished Master's thesis investigating meanings of sustainability in Svalbard tourism, Dvorak (2019) addresses the feeling of not being welcome in Longyearbyen as a non-Norwegian guide: 'The Norwegian residents actually don't like us, but they tolerate us because nobody else wants to work under such conditions as we foreign guides do' (quote cited in Dvorak 2019: 25).

During an online meeting organised by the Ministry of Justice and Public Security in May 2021 as part of the hearing of the new Tourism Regulations (*Turistforskriften*) likely to enter into force in 2023, the leader of the Svalbard Guide Association identified the international guides as being those 'at the bottom of the food chain'. Another guide appealed:

> Svalbard attracts lots of both international tourists and operators. [...] In my opinion, it sends a very unfortunate and not very inclusive signal to send out the meeting information only in Norwegian and to mostly Norwegian actors. The hearing itself is now mostly only in Norwegian. It simply shows little recognition of the knowledge of all these foreign operators and individuals. [...] We guides love Svalbard, nature, the environment and wildlife. As much as all of you, if not more! We care for Svalbard and as true ambassadors we want to secure the environment on Svalbard for future generations. [...] It is extremely important that foreign guides can obtain these future certifications, which will be required to be a guide on Svalbard, in a practically feasible and affordable way. [...] Previous austerity and changes often came abruptly, surprisingly and focused on areas that felt strange to many involved who know the actual challenges in the field.

These examples illustrate the major clash between how Svalbard is being sold as a place transitioning from an unsustainable past to a more sustainable future while traces of extractivism remain alongside increasing pressure with a nationalistic undertone. They also relate back to Toni's comment about what does not work well: tour guides, key mediators of how people travelling to Svalbard make sense out of the place's many stories, including the one of climate change, are an international, underprivileged and politically unrepresented group of people on the margins of the transient community, where tourists arrive with racist prejudices and colonial ideas about Svalbard: 'Norwegian clients complain that the tour guide is not Norwegian. Not that I don't speak the language. They think it's only Norwegians who can survive the cold. They don't like that in their place a foreigner should guide them.' Another interlocutor remembered a highly qualified tour guide from West Africa who had to leave because clients reported they were unhappy with 'a driver' guiding them in the 'Arctic wilderness', and a local Filipino man who finished the basic guiding course but never found a job because he was short, thin and Asian-looking, which 'just doesn't fit the image of an Arctic nature guide'. Such issues are hardly ever included when the sustainability of Svalbard tourism is discussed; environmental and economic questions take precedence.

The lived experience of the tour guides demonstrates the need for a much stronger focus on a workers' rights agenda, including disrupting labour hierarchies and divisions based on gender, race, and nationality (Christian 2017). But exploring the struggles of tour guides in Longyearbyen does more than that. It turns attention to 'what stories can – and cannot – be told and how' (Ren et al. 2021: 117) about Longyearbyen, and what it is like to live there amidst rapid changes. The third part of the book zooms in even more on the 'little people' of superdiverse Longyearbyen, asking how the trouble with local community relates to disempowerment, nationalism and a culture of denial.

PART III

Disempowered Communities

7

The Trouble with Local Community

'Norway must have presence in Svalbard. It can be coal mining, it can be making jam. It doesn't matter. Norway will pay. And the people there ... they have no meaning.' The straightforward wording of the high-ranking stakeholder representative I talked to in 2022 about Longyearbyen struck me. On earlier occasions, my conversation partner had claimed that he is often 'sitting in the room when decisions are being made about Svalbard'. I had no reason to doubt his words.

To consider human lives meaningless is an alien take in today's anthropology. In ethnographic fieldwork meanings unfold all over the place. When I arrived in Svalbard, I noticed two spaces in the central public space of the town that embody the homogeneous post-war past of the Norwegian Longyearbyen. One of them is the statue of *The Miner*, encountered in Chapter 4. Just a few tens of metres from *The Miner*, active and retired miners and their friends gather on Tuesdays at 10 a.m. around a table in the town's most popular café, Fruene (The Misses). When the café was renovated after 16 years of use in 2020, the owner deliberately preserved the table of the regular customers who represent a connection with the town's past. Interestingly enough, the 'grumpy old men', as some refer to them with love and esteem, decided to test other spots in the café as well and eventually found a table that suits best their need to drink coffee, watch people passing by and chat in the protective niche where they are respected. When I first came to the 'old' Fruene in July 2017, I watched elderly men sitting around a table they apparently occupied with great confidence, speaking a language that was then incomprehensible to me. In my fieldwork diary, there is a note about a man I estimated to be in his seventies, in a thick woollen jumper, with white and somewhat untidy hair, a long white beard and piercing blue eyes. I noted: 'This man I would like to talk to. Will I ever manage it?' When I moved to Longyearbyen in February 2019, he was no longer to be found at Fruene, but I could meet others, presumably his companions. I certainly did not see them as meaningless; quite the contrary. I saw the geopolitical backdrop of the place, its history and historical legacies as chapters of a multi-layered story that intrigued me, but the lives of people I could meet and talk to lay at the core of my work.

Previous studies show that Longyearbyen is Norway's bastion of national presence (Roberts and Paglia 2016; Pedersen 2017), a place of great geopolitical significance but also high population turnover. Neither of these aspects is new, but during recent decades, global developments have amplified both of them. The interest in the Arctic, including Svalbard, has grown, and travelling and migrating to Svalbard has become easier and more attractive. There is a shared perception of these recent developments circulating among people in Longyearbyen that can be summarised as 'A lot is happening and it's all happening too fast.' I cannot count the number of times I have encountered this framing and noted it in my fieldwork diary, be it at political meetings, at demonstrations, on social media, in informal conversations or during formal interviews. Many experience the overwhelming impacts of disempowering change, which Eriksen and Schober (2016) describe in their book on what living in an overheated and globalised world is like: people feel they 'did not initiate the changes themselves' and that 'they were not even asked by anybody in power for their opinion' (Eriksen and Schober 2016: 3).

> We need something to gather around. When we light up the Christmas tree, we walk around [it]. Everyone is part of that because everyone has a sort of a connection to Christmas. [...] The Christmas tree gives us all the good feeling we need to find. [...] We need something to hang all these beautiful decorations on, that everyone can contribute to. Perhaps we should start making ornaments for the Christmas tree. All of us. But the ornaments could be things, tasks. Yeah. Engagement. I don't know.

Over another cup of coffee at Fruene, about a month after I moved to Longyearbyen, Sunniva shared her metaphor of a missing jointly decorated Christmas tree that would enable the people in town to perform as a community. There is a Norwegian tradition of dancing together around a newly lit Christmas tree observed at *torget* in Longyearbyen. Sunniva was not herself convinced by the metaphor. It is not true that everybody shows up at such gatherings. Having lived in Longyearbyen for over 20 years, Sunniva was looking for an image to illustrate the growing need to strengthen social cohesion in the transient settlement.

The Christmas tree metaphor is a somewhat linear, hierarchical one. People are instrumentalised as decoration producers, expected to contribute to something they may not even identify with. Who gets to hang their ornament on the visible top, and whose engagement gets less privileged spots? In this chapter, I trace what happens when a settlement that is quite homogeneous goes global and when differences become more pronounced. The dancing around the tree, the winding movement of people expected to be close to each other and to celebrate their community's cohesion is an image evoking nostalgia for times when difference and even absence from the life of the community were less common. There is a sense of a particularly subtle, probably also unintended violence in Sunniva's pressing those who do not conform to feel obliged to engage, show up and perform the dance.

There is something paradoxical about the discourse of the 'local community' in Longyearbyen, a clash between what is desired from above and what people long for from below. Let us follow the twists of the *ouroboros* that revisits the unspoken and the unheard, disrupting the narrative of miners being replaced by researchers, all of them preferably Norwegian. The third part of the book dwells on perceptions of how change impacts life in town, the population's texture and the socialities emerging from the change. What is it like to live in a place where people repeat 'We are all here for a reason,' reminding themselves of the place's geopolitical purpose? And how do processes intrinsically bound to globalisation, such as higher mobility, relate to the reasoning about 'meaningless humans'? People settle in Longyearbyen for a wide variety of reasons and ascribe to it various meanings. However, some life stories fit in the discourse of Longyearbyen depicted as a Norwegian family community dancing around a Christmas tree more than others. With increasing heterogeneity, the margins of inclusion are shrinking. While the geopolitical scope has remained intact for over 100 years, the settlement itself has been changing and the changes make the tensions express themselves in new ways.

How Unique and How Norwegian is Longyearbyen?

Longyearbyen stands out as a settlement lacking an Indigenous population, since it was created for the purpose of resource extraction and sustained for the purpose of exercising presence in order to claim continuous sovereignty. The former disqualifies Longyearbyen from comparison with places inhabited by people who, for generations, have developed a strong sense of belonging and who also have legally acknowledged rights related to the territory. The notion of people's temporary stay, expressed in exclamations such as 'We are all tourists here in one way or another' or 'This is such a limbo place' became a recurring theme in my ethnography. Instead of an Indigenous population, there is a 'local' one. A feature that makes Longyearbyen unique among other extractivist settler communities is the comparatively low legal hurdles people face when relocating there (no visa and no work permit needed), thanks to the terms of the Svalbard Treaty (see also Brode-Roger 2023).

Any comparison made between Longyearbyen and other communities is thus made with caution and with two aims: to explore meanings the town's inhabitants ascribe to how communities work (or should work) and showing similarities to other communities while bearing in mind the place's uniqueness.

For the sake of the former, Scandinavian (or Nordic) and, more specifically, Norwegian sociality (Bruun et al. 2011) is of particular relevance given the town's history, frameworks of governance and the population's majority. On a few occasions, my Norwegian interlocutors independently described Longyearbyen as un-Norwegian (*unorsk*). When prompted to elaborate, they would point to the striking difference between how hierarchies are pronounced, growing and even fostered in Longyearbyen, and the egalitarianism, solidarity and commonality that are idealised aspects of Norwegian community life (Bruun et al. 2011).

In the international context, including academia, Norway is often seen as a 'a homogeneous, tolerant, anti-racist, and peace-loving society' (Gullestad 2002: 59), in addition to being a climate do-gooder (Anker 2020). The multi-layered paradox of Svalbard I am trying to unpack in this book challenges this representation. As Abram (2018: 90) points out:

> it is the idea of a rich yet egalitarian nation which has positioned Norway as a most exotic land on the European periphery, often invoked [...] as a kind of welfare paradise, courtesy of oil income, high taxation and its egalitarian culture, and, as such, as a kind of exception to the normal rules of European economics and politics. On the basis of this egalitarian culture, Norway can be seen as both a source of inspiration for other countries and as an impossible ideal.

The rhetoric the Norwegian state uses to shape the future of Longyearbyen is specific given the peculiarities of Svalbard's territorial regime, but it cannot be understood without seeing how it is rooted in Norway's image of the Self. In Longyearbyen, the welfare system is available only to an exclusive layer in the society (see Chapter 8 for more detail) and the taxation is low. This reinforces the argument that only those who contribute to the dance around the tree may benefit from the state's generosity to boost Longyearbyen financially. As many Norwegian anthropologists have documented (e.g. Eriksen 1993; Gullestad 2002; Karlsen 2018), much of the discourse representing Norway as 'the best country in the world' can be read as 'nationalism in action' (Abram 2018: 90). The way it manifests in Longyearbyen is both typical and unique in the Norwegian context. In a way, there is a paradox both in realising that the way in which Longyearbyen is un-Norwegian (*unorsk*) is actually typically Norwegian (*typisk norsk*), and that this 'normalisation' of Longyearbyen was accelerated by economic plans the Norwegian state drafted and put into practice itself. The state wants a showroom of ideal Norwegianness. What it gets is a Norwegian reality in a High Arctic setting with 'vulnerable nature' impacted by climate change in the background.

Also the unease with which difference is handled in Norwegian communities is apparent in the changing Longyearbyen. In his classical study of class differences in Norway, Barnes (1954: 50) claims that:

> individual goals must be attained through socially approved processes, and as far as possible the illusion must be maintained that each individual is acting only in the best interests of the community. As far as possible, that is, the group must appear united.

While Longyearbyen in an odd way combines the features of a rural small-scale society, where social networks are tightly knit, with the mesh of an urban society, the town's legacy of having been a Norwegian company town lives on in today's mourning of a united 'community'.

It is, however, useful also to see how Longyearbyen resembles places beyond the Norwegian context, founded on the premises of extractivism and with a high turnover of newcomers attracted by globalised job opportunities. In the Latin American context, Stensrud (2016) writes about Majes as a place that is special because of the 'non-existence of a pre-existing population, and the diversity of its population, coming as it has from all places and classes' (2016: 75). Social differentiations and hierarchies are emerging where scale-related tensions are palpable between the lived experience of individuals, local issues, regional and national affairs. A regionally closer example is Kiruna in northern Sweden, a heterogeneous 'masculine working-class town with ethnic tensions' (Granås 2012: 135) with industrial legacies and a normative ideal of outdoor life. When studying how the townspeople handle their relationship to the place, the future of which is largely shaped elsewhere, Granås criticises scholars victimising the inhabitants, representing them as disempowered and compromising their agency that might rather lie in pragmatism.

Ambiguity and a pragmatic attitude are explored in studies of communities based on extractivism beyond the circumpolar context; an example is Laastad's (2021) study of how 'processes and conditions that include the local scale reshape political geographies of extraction' (2021: n.p.). In her case in Peru, she documents the 'uneasy coexistence' with local impacts and dependencies on decision-making processes that are co-shaped by national and economic interests in a global context.

In the circumpolar setting, Longyearbyen bears some similar traces to communities in Greenland, northern Canada or northern Fennoscandia struggling with the pains of boom and bust cycles. Shared elements include replacing or combining mining with tourism and research, experiencing the ambiguous impacts of both environmental destruction and protection, and other issues where rationality mixes with emotions (Sörlin 2021). Threatened Indigenous livelihoods due to climate change impacts, combined with globalised markets (a problem often documented in Arctic anthropology, as shown in e.g. Hastrup 2019), are not an issue in Longyearbyen as there is no traditional way of sustaining life locally without support from the mainland. All these places in the Arctic are redefining their futures in an overheated world where a new extractivist paradigm has been emerging (Sörlin 2023).

The Arctic is home to about 4 million people today and, apart from the Russian Arctic, the population of the area is expected to grow (EEA 2017). In the case of Longyearbyen, the speeding up of two trends during the past two decades counteracts the Norwegian governmental strategy for Svalbard – namely that Longyearbyen is expected to be (a) a Norwegian community (*samfunn*) and (b) population growth is discouraged (Norwegian Ministry of Justice and Public Security 2015–2016). Both the increased level of the population's heterogeneity and steep population growth (doubling since the 1990s) are typical for boomtowns (Eriksen 2018).

Only since the early 2000s is it possible to situate Longyearbyen in the context of multilingual and international immigrant communities. Mobility and community have become conceptual twins in the twenty-first century in Africa as well as the Arctic, when 'human mobility engenders socio-political interactions [...] that form and reshape the meaning and boundaries of community' (Landau and Bakewell 2018: 3). In some aspects, Longyearbyen can be seen as a rural locale where people appreciate short distances and a less stressful life, an 'elastic' place constituted by mobility (not only the mobility associated with tourism), with numerous transnational migrants (Milbourne and Kitchen 2014). In the Milbourne and Kitchen study, situated in rural Wales, the authors summarise the perception of changing mobility patterns as follows:

> The movements of different groups into, out of and through rural spaces and places have not only altered the demographic profile of the study communities but also impacted on their local sociocultural and linguistic composition. Indeed, some long-term residents discussed these migratory processes as destroying place, undermining community cohesion and damaging long-established cultural norms. (Milbourne and Kitchen 2014: 334)

In Longyearbyen, people with a longer personal history on the island talk about an obvious change in the demographic texture, sociocultural and linguistic outlook, and community cohesion. It is possible to draw a parallel with Nordic rural places shaped by migration from all around the world, with ambiguous implications: it may lead to increased municipal expenses, tensions between the established population and the newcomers because of fear of competition for housing and/or jobs, or concerns that immigration may challenge the local culture. On the other hand, immigration contributes to the local labour market, and helps to stabilise or increase local population levels (Søholt et al. 2018: 220).

Although such an approach to migration bears instrumental traces, I met reflections on the majority of the impacts listed in the previous paragraph in conversations with people in Longyearbyen. There is only one exception. It is true that a constant influx of workers (smoothed by the legal repercussions of Svalbard Treaty that keep the area out of Schengen and visa-free) is necessary for the labour market, both in niches typically occupied by Norwegian or non-Norwegian employees. Yet in the political and media discourse 'local population figures' are presented in a normative manner; it matters whether the new settlers are Norwegian, or not. According to the governmental strategy for Longyearbyen, keeping Longyearbyen populated is not enough; the ratio of Norwegians to non-Norwegians should increase. Migration from outside Norway is thus seen as a problem, framed as a security threat for Norwegian sovereignty (Pedersen 2016). In their study of Nordic (Danish, Swedish and Norwegian) rural communities with high rates of foreign immigration, Søholt et al. (2018) explore the link between resilience and retention/ receptiveness of 'local elites' towards international migrants, be it 'short- and long-term labour immigrants, circular and seasonal workers, as well as refugees, and family immigrants' (2018: 221). Except for refugees, given the specific territorial regime of Svalbard, all the listed groups are part of the 'throwntogetherness' (Massey 1994, 2004) of today's Longyearbyen. Its immigrant population is three times larger than an average Norwegian peripheral locale (38 per cent vs. 11–12 per cent). Many patterns are similar; foreign immigrants tend to live in smaller houses, they are often employed in booming industries such as construction or tourism, and they accept jobs the majority would not take. Søholt et al. (2018: 227) conclude:

> Immigrants' inclusion in the labour market as such may contribute to local economic development, whereas ignorance and/or lack of social inclusion might simultaneously produce a segmented labour market and parallel social communities. Processes of exclusion/inclusion and retention/receptiveness that produce new, undesirable inequalities might imply less stability, and deprive the community of the potential of all residents participating in co-producing the common and desirable future of the rural place.

To 'Communitify' Instead of 'the Community': Focusing on the (Un)Becoming

> Just as Arctic settlements have historically tended to be temporary due to their reliance on fluctuating resources, contemporary Arctic communities related to extractive industries continue this tradition. [...] Local communities came to face existential challenges [...], and they chose to respond to these challenges by means of strategies of communitification. I understand the verb 'to communitify' as the discursive act of pulling people together into a group, articulating boundaries around them, and attributing them certain emotions, behaviours, motives or traits. I observe the uses of the concept of community, its transformations in processes of communitification, and the agencies deriving from its being articulated either from the inside out, by the group itself, or

from the outside in, by authorities, politicians or other actors in the surrounding society. (Jørgensen 2019: 1–2)

I apply the idea of communitification as a process, a dynamic strategy involving values and emotions, used by Jørgensen in her comparative study of Qullissat in Greenland and the village of Sakajärvi in Sweden. Jørgensen is aware of the illusionary imaginary of Arctic communities as isolated and 'local', while in reality they 'encompass ethnically mixed, highly mobile, and composite identities,' (2019: 2) which characterises Longyearbyen accurately. As I will show in my empirical material, 'local community' (*lokalsamfunn*) in Longyearbyen appears almost as a 'mysterious character', something that everybody talks about, that is mentioned in all governmental and strategic documents and in the media, but slips away from every effort to capture it in words (Norwegian or English).

The aspect of change in time is highly pronounced here, with long-term residents witnessing a change in what the term 'community' would mean to them as they watch the highly transient town transform into a diverse and multi-ethnic place and with the texture of the population no longer legible. As Sverre, born in Longyearbyen and living here throughout his life, commented:

Sverre: When I go to a pub I kind of expect people to know who I am. But they don't any more. [...] I guess this is part of the ownership towards this place. I grew up here, I have been living here for so many years.... This is my home town. But it's not any more ...

Zdenka: So when you hear the term *lokalsamfunn*, what do you imagine?

Sverre: That's difficult. Perhaps I go back to the time when I went to high school. I think of those people. But they are not here any more. It's a difficult question. I don't know what to answer. We are in constant change. I don't know. [...] I mean, it is everybody but ... is it?

The normative term of (local) 'community' as a poor fit for processes that entangle people and places in a mobile, volatile and overheated world bounces back in reflections such as Sverre's. Longyearbyen's social memory, mourned by Sverre, is something typical for communities; a collective past that 'draws on both the memories and life histories of its individual members, as well as the social context that they experienced together' (Lyons 2010: 25). It refers to emotions, mostly nostalgia for something known, legible and safe.

Anthropologically and sociologically intensely researched, 'community' (together with 'culture' or 'identity') is among terms many have investigated and argued over. The understanding has shifted together with changing patterns traceable in human (and eventually more-than-human) lifeworlds. The way Cohen (1985) or Anderson (2016 [1983]) approached community, as an imagined entity symbolically constructed, 'a system of values, norms, and moral codes which provides a sense of identity within a bounded whole to its members' (Editor's foreword in Cohen 1985: 9), was revised by the turn of the millennium by thinkers such as Bauman (2001), who still saw community as something positive, but also paid attention to the fluid and non-static process of boundary making. Changing patterns in mobility and identity-making challenged earlier theories. The 'trouble' with community – still sticking around as something desirable and homogeneous – was recognised by authors such as Amit and Rapport (2002) or Creed (2006). The self-explanatory potential of the term is limited since we know that 'different inhabitants have different opinions, conflicting agendas, and different connections to the centers of power' (Crate 2008: 570). The term 'communitification', applied by Jørgensen (2019), allows for focus on processuality and relationality instead of a static definition essentialising such an incoherent group of people as Longyearbyen's inhabitants. It might help to use a verb instead of nouns, a word that captures the potentiality of a political mobilisation instead of blurred yet powerful labels.

The Norwegian word samfunn can bear meanings that English (or German) translate in a more nuanced way; both 'society' and 'community' (in the sense of Tönnies [2001] Gemeinschaft and Gesellschaft) can be intended. In addition, the political discourse of lokalsamfunn (Norwegian Ministry of Justice and Public Security 2015–2016; Longyearbyen Lokalstyre 2013) can be read mostly in the sense of a settlement constantly being re-populated with citizens of Norway, who are provided with basic services necessary for a middle-term stay. Though in the updated version of Lokalsamfunnsplan 2023–2033 (Longyearbyen Lokalstyre 2022), the word 'settlement' (bosetning) is used only once, the term samfunn (or lokalsamfunn) appears 112 times. This clearly directs the meaning of samfunn towards community – not a neutral settlement, but a place the meaning of which is created through shared values: 'Longyearbyen is a safe and inclusive local community whose inhabitants feel a sense of belonging and accept responsibility, regardless of cultural, social or economic conditions' (Longyearbyen Lokalstyre, 2022: 17).

My data documents a strong wish to conjure a sense of group identity allowing Longyearbyen residents to gain a 'claim on the future'

(Schulz-Forberg 2013, as quoted in Jørgensen 2019). The lack of such a claim is stressful when facing insecurities such as climate change impacts, strict environmental and tourism regulations, politicised housing, and a lack of services accommodating the needs of those who do not fit the imagined identity of Longyearbyen as a Norwegian settlement consisting of healthy visitors (*besøkende*), as the governmental documents understand anybody living in Longyearbyen to be. Using Jørgensen's approach, I intend to show in this chapter that, unlike other Arctic communities that at least partly succeed in mobilising for the sake of communitification, allowing them to play a more active role in their home's future-making, a similar process is hindered in Longyearbyen.

Living in a Samfunn without Fellesskap

> While in much everyday language 'community' appears as a descriptive term, in the present contexts, people use communitification as a strategic tool in the negotiation of rights and ownership, and an instrument in their quests towards certain desired futures. (Jørgensen 2019: 1).

> One of the overriding objectives of the Svalbard policy is maintaining Norwegian communities in the archipelago. This objective is achieved through the family-oriented community life in Longyearbyen. Longyearbyen is not a cradle-to-grave community, and there are clear limits to the services that should be made available for residents of the community. This is reflected in the archipelago's low level of taxation and the fact that the Norwegian Immigration Act does not apply here. The government's aim is for Longyearbyen to remain a viable local community that is attractive to families and helps to achieve and sustain the overriding objectives of the Svalbard policy (Norwegian Ministry of Justice and Public Security 2015–2016: 39).

The 'desired futures' of Svalbard are being designed elsewhere; the very purpose of communitification – negotiation of rights and ownership – is tricky where national interests have an automatic override. That is by no means new in Longyearbyen. Yet some who settle down in Longyearbyen dislike being seen as 'puppets in a geopolitical theatre' and they also notice change over time.

Olaf has been living in Svalbard for more than 40 years and witnessed the evolution from a company town to what it is today; some call it a 'state town', some a 'tourist town', some a 'melting pot', some a 'testination'. Given its size and geographical location, the diversity of Longyearbyen's inhabitants is extreme and has been increasing fast since the last survey with a comparable focus (Moxnes 2008).

> *Olaf*: I perceive the development in the community (*samfunn*) here as something positive, in contrast to some others. [...] From a Norwegian community (*samfunn*), we have become multicultural. It's more an enrichment than something negative. I think it's stupid that we focus on it so much. I think it's just individual political powers, both locally and on the mainland, that use it. [...] Unfortunately I see also here in Svalbard that people have become more selfish. Egoism rules in the whole world, the community (*fellesskapet*) is on the way down. And I think that if you don't embrace the community (*hvis du ikke favner på fellesskapet*) in such a small community (*samfunn*) like this one then we all lose. [...] I don't even understand why some say this should be a Norwegian community (*samfunn*). It's correct that Norway has sovereignty, but that's where the line goes. I'd rather say we could be a pioneering place (*foregangssted*) to show that here we can manage.

When we compare the formulations from the Norwegian government's White Paper with Olaf's reasoning, the angle of the politics of translation might prove helpful. When Joks et al. (2020) investigate (mis)translations of Sámi words weaving into Norwegian and English, they show how the colonial mindset is mirrored in the way we go about using words. Despite the irrelevance of the Indigenous context, Svalbard politics demarcates the borders within which only what people call a 'fake' or 'synthetic community' is allowed to develop. *Samfunn* here means a settlement empty of locally shared values. For Olaf, a *samfunn* without *fellesskap* is dysfunctional, and he claims the communitification potential (thus, filling up something empty with emotions and values) has deteriorated during his lifetime.

The multicultural aspect of Longyearbyen, typical for the last two decades, accompanies the diversified job market. However, not all sectors employ equally across nationalities. Local authorities (such as the Governor of Svalbard and LL), Store Norske, Statsbygg, SVALSAT (the satellite station operated by Kongsberg Satellite Services, a joint venture between Kongsberg Defence and Aerospace and the Norwegian Space Centre), Avinor (a state-owned company that operates the airport),

Lufttransport (a Norwegian helicopter and fixed-wing airline), LNS (a Norwegian family-owned enterprise in the construction industry), Telenor (a Norwegian majority state-owned multinational telecommunications company) and others have a vast majority of Norwegian employees. Also almost exclusively Norwegians are employees of other institutions, such as the hospital, bank, post office, *Svalbardposten* and tax office. A recent phenomenon is moving offices of entities that have little or no relevance for Svalbard to Longyearbyen, such as that of *Forbrukerrådet* (the Norwegian Consumer Council, which in 2021 transformed into a local branch of *Forsbrukertilsynet* – the Norwegian Consumer Ombudsman) or *Norsk Helsenett* (Norwegian Health Net). The goal is to establish more jobs that will most likely attract Norwegian employees who will only stay for a few years and leave again (Forbrukerradet.no 2017; Nhn.no. 2020). Some of these institutions moved in 2022 right into the centre at *torget*, into a building that now bears the name 'the State's House' (*Statens hus*) – a sign of its presence that cannot be missed. The *Folkehøyskole* (Folk High School), established in 2019, brings over 100 students to the island every year, the vast majority Norwegian, also with an explicit aim of increasing the Norwegian ratio in the population.

These jobs are seen as 'safe' in the sense of offering fair and legal working conditions, job-related benefits, adequate housing and health insurance, which also makes them attractive for non-Norwegian applicants. Getting one is not easy. One of my non-Norwegian participants, who was working in the public service sector, recalled how surprised they were to be offered one of these jobs, expecting a Norwegian candidate to receive preference. Another non-Norwegian acquaintance shared that it was unofficially communicated to them that applying for state-funded jobs without a Norwegian passport is a waste of time. It is the other distinct sectors – most importantly tourism, the service industry, research and freelance writers, journalists, editors, graphic designers and so on – that are essentially international arenas. Since the 1990s, the job market in Longyearbyen has been ethnicised, which is a pattern visible also in mainland Norway.

People staying for several decades, such as Olaf or Wenche, who are both part of multigenerational Svalbard-connected families, are aware of the inherently transient character of Longyearbyen. But they see something is happening to the community's cohesion and agency potential.

Olaf: What was perhaps a bit better earlier is cohesion (*samholdet*). People lived here longer and knew each other better. [...] Earlier everybody knew almost everybody. Today when people come we know they will probably just stay for a year or two. When you talk to them they are here just to experience Svalbard, a short period in their lives, to have it on their CV, often to get a better job later, and then you just don't have it in you to engage. To have good friends, that requires something from both parties. [...] It's very good there is a new generation but we also have to strive for continuity. [...] There aren't people with knowledge any longer. [...] We who have lived here long, we have a totally different attitude towards Svalbard than the new ones. Svalbard lies deep in our hearts (*ligger i hjerteroter våres*).

Olaf is not complaining about non-Norwegians; he is worried about the impact of the population's fluidity, lack of attachment to place, and extractive sense of place. More people in town means that even with a similar level of turnover more people come and leave. Those living in Longyearbyen see as the biggest problems the lack of continuity and loss of knowledge caused by the turnover, as well as growing differences and the striking political under-representation of certain groups. Olaf's feeling that continuity and engagement in shaping Longyearbyen's present and future are weakening is shared by all the Svalbard old-timers I met during my fieldwork, but it also resembles the feeling of Susan, a European 'newcomer' employed in tourism:

Susan: Even in three years you could see the the change. [...] Before it was a lot of people ... well, a lot ... it was 500, 1,000, anyway, but they were moving every year, second year, third year maximum. But then [...] in the 1970s, 1980s, they started to stay. To make a real community. And then they stayed for a long time. They got involved with the place, they loved it, cared about it. But now it's going back again. For me, that's definitely because of tourism because that's something that works in seasons and then you have seasonal workers. Even the building [industry] is seasonal so you have a lot of seasonal workers in the winter to build. [...] And all these people, they don't care about this place.

In Susan's memory leaning on the memory of others, a sort of a romanticised past is being evoked, a time when there was 'a real community' of Longyearbyen. She is critical about the industry that enables her to make a living. What is noteworthy about the perceptions of people whose room for manoeuvre is very different in terms of how established/precarious their lives in Longyearbyen are is that they identify care and engagement with equal urgency as values that hold the 'community' together. Since the turn of the millennium, local governance has shifted towards a place-specific form of local democracy, but the town has also become more diverse and cosmopolitan. Both changes bring challenges that – unless tackled – hinder communitification:

> *Wenche*: When I came here in [the 1980s] it was still quite usual that most of the state jobs, apart from Store Norske which actually wasn't a typical state employer, were tenure positions so there was a natural turnover. So already then there was a gap between those who worked for the state and turned in intervals of 4, 5, 6 years, and those who worked for Store Norske and had stable jobs. [...] And then there was another group called 'the others' (*andre*), who were basically the spouses or family. Because by then there was no private market or entrepreneurship. There was no local democracy like today, with political decisions, but there was the Svalbard Advisory Board (*Svalbardrådet*) with representatives of all the groups. So it mirrored kind of the texture of the population, with the state, Store Norske, and 'the others'. [...] People were often very much defined by their jobs. [...] Longyearbyen has always been a very class-divided society, defined by professional identity which impacted job-related benefits. [...] But I didn't see as if ... yes, you were placed according to your job, but in the social, cultural life, in your free time, in the streets you could interact (*samhandle*) without any separation (*skille*) or limit (*grense*).

Wenche's perception through the perspective of several decades in Svalbard is close to the interpretation of Susan. Interaction is complicated by the usual class divide, newly accompanied by the split between people with and without a Norwegian passport (often combined with a gap in language competence), and the intensified transience:

> *Susan*: I see three kinds of segregation. For me there is definitely something about where you come from because, of course, some communities are big enough to just stay together, and you have these groups by country or group of countries. [...] The second is definitely about work. People bond so much in their work. And even if they can be in clubs and other activities and cross, you still have people, for example, working in offices so having the same time [schedule] [...] then waiters

and people working in kitchens are a lot together, the guides are also together because, of course, they have a crazy schedule and basically when they are off they just want to sleep and share their experience talking to other guides, you know. [...] And the third kind of segregation, that's your Svalbard age. There is not too much interaction between these groups and you can feel it.

The hundreds of people applying for jobs within the tourism and the service industries are exposed to the impacts of the usual mechanisms of costs externalisation, combined with the governance and legal legacies of the company town. Their employers are not legally bound to provide housing or language courses. There is no valid legal framework granting them a safety net when it comes to health care or unemployment aid. Only a few job benefits accompany the often seasonal, short-term contracts. People working in the sector are caught in the dead end of a contradictory element in the Norwegian strategy for Svalbard: honour the Svalbard Treaty, support tourism and keep up the ratio of Norwegian inhabitants of Longyearbyen (compare to Pedersen 2017).

In an informal conversation with a non-Norwegian entrepreneur, they recalled a situation when they visited the office of Innovation Norway, a state-owned company and a national development bank that at that time still had a branch in Longyearbyen. The entrepreneur was encouraged to apply for a grant and start up the business, but the Innovation Norway officer 'kindly reminded [them] that it is desirable to employ Norwegians in the company, if it ever grows and [we] need more people.' The encouragement is perceived as embarrassing, as another participant confirmed – a Norwegian who previously worked in the tourism sector and moved to a state job afterward. They told me about a meeting the Governor's Office arranged for tourism managers in the late 2010s, when the presentation about new possibilities opening up for tourism concluded with a reminder that 'Now, it is up to you guys to employ Norwegians here.' My participant reacted with astonishment: 'But you know what, this is private business!' Another participant, a local businessperson in the service industry sector, wanted to comply but did not succeed: 'I have tried to employ Norwegians but they are unusable (*ubrukbar*).'

The term 'unusable' has interesting connotations. First, it addresses the core premises of the business, which is profitable only if 'human resources' are 'used' efficiently. Efficiency does not only include work ethos, diligence and accuracy; it might also consist of willingness to accept suboptimal

work conditions when it comes to working hours, physical and psychological fatigue, housing or salary (or all of that). The somewhat cynical
observation of my participant also points to the fact that Norwegian job
applicants are more likely to speak up if they have a good reason to believe
conditions are not acceptable, and they are able to organise and fight for
their rights, like Ole from Chapter 6. Norwegian guides I have spoken to
confirm this univocally and mastering the language of power – Norwegian – is a factor that should not be underestimated here. The language
barrier that delineates spaces of information accessibility, but also agency
(most urgently but not exclusively among Thai and Filipinx migrants, see
Chapter 8) is perceived as a political tool.

> *Simon*: Longyearbyen is getting way more political. [It] is getting way
> more changed from the outside and it is getting far more aggressive
> to non-Norwegians. For example, there was a language course up here
> when I first came. [...] I had an active Facebook battle with some of the
> people here in town because they closed down the language course. And
> then I was literally told by people that it's bloody foreigners who had
> better take care of our language skills ourselves if we want to integrate
> with Norwegian society. And that's a bit of a weird statement because
> especially language, of course [...] should be accessible for people to
> integrate. [...] The second example would be, I'm guessing, the housing
> crisis. [...] There is a solution, but people don't want to do the solution
> because it doesn't fit the political motivation. [...] So there's a lot of
> these small hints that give you those kind of views.

When Simon is trying to make sense of the 'hints' he is getting from
different directions, he does so against the backdrop of his previous life
elsewhere (in different countries on different continents), where he also
experienced racism and discrimination. While Svalbard might look like
a utopia from the outside – an area in the far north where everybody is
welcome – Simon realised over time the illusion is manifesting itself in
an ever harsher way. What he would expect to be obvious issues to tackle
politically have become normalised. Some of my interlocutors confessed
that when they left Svalbard, they felt relieved of this politically accepted
pressure. While the situation of people employed in science, such as
Simon, and those employed in tourism and the service industry differs
in terms of housing (with researchers being granted housing through
their employer, while employees in the latter group often must rely on

the volatile and expensive private market), they share the same difficulty in learning Norwegian. According to Susan, it hinders her involvement in shaping the town's future from within: 'Perhaps I am not so much part of the community as I could be if I spoke the language. Because I am engaged. In transition and social responsibility, you know.' When asked about what she associates with the term 'local community' in case of Long-yearbyen, she claimed:

> Susan: The nature is harsh. And when the conditions around are already difficult, then people are usually more sticking together. I was expect-ing more bonds. Inside these groups, the bonds are very strong, but outside.... And now even more because people don't stay so you don't have this mutual assistance and help. Caring. If it makes sense.

The issue of solidarity was brought up by numerous participants, espe-cially in connection to the state of emergency caused by the avalanche in 2015 that works as a rupture in the collective memory of the place (see Chapter 2). Yet once again, change is palpable in the short period of several years. Simon explains:

> Simon: I think if shit really hits the fan, then that is when the local community steps up. So, for example, during this avalanche that hit the houses and killed two people. That was really when you said, wow, this is how the community works, but that was [...] five years ago. [...] But since then, there's actually been some quite big changes in the town because there's a lot of coal miners that have left. So the entire tradi-tional sense of what the town was is no longer really the same. And tourism is taking a far bigger part, and I'm not sure whether tourism here is sustainable. [...] I do think maybe having more industry back up here, like traditional industry, whether it's coal mining or extend-ing KSAT or whatever, that might help [to make] these people stay here hopefully for 5 or 10 years rather than leaving every year. And I do think that bringing people to work at, for example, LL or the Gover-nor's for just a limited time of three or four years is [...] not enough for a long-term vision. [...] You have a lot of people living here that actually want the best for the place, but [others] that actually just listen to Oslo. And if Oslo says 'you do this' they will do that because in three or four years they'll be on a different duty station anyway. That is maybe not the most healthy for a small little settlement like Longyearbyen. But I think

[...] it's not really here for Longyearbyen people. [...] Longyearbyen is a geopolitical pressure point for Oslo.

The perception of people like Susan and Simon, working in tourism and science, fighting the language barrier and practices that on the individual level are experienced as discriminating for higher political purposes, does not differ from the perception of long-term Norwegian residents who have lived through the turbulent era since the 1980s, such as Wenche:

Wenche: I'm afraid parallel societies (*paralelle samfunn*) have been established in Longyearbyen. [...] If language is the entrance ticket to the community then quite a big part of the population doesn't have that competence. [...] I think it's quite dangerous for Longyearbyen to have such parallel communities because sure, we can live next to each other in many different ways, but if there are too big contrasts (*motset-ninger*) a conflict will come at some point. Take housing. Or salary and working conditions. Somebody is within (*innafor*) a system that is regulated, and somebody is outside the system (*utenfor*). [...] There are already well-developed parallel communities and a moment will come when you either must address it to do something about it, or it will keep developing and it will get more difficult to bring them together.

The existence of layers in society is not remarkable, even in the supposedly egalitarian Scandinavian context (Gullestad 1984; Bruun et al. 2011). Neither is the issue of language a simple one. In their study of immigrants' perception of learning Icelandic, Skaptadóttir and Innes (2017) discuss whether learning the language of the majority for an instrumental purpose (e.g. to have access to information or socialise more easily) also means a gateway to inclusion. Their results instead show language is a tool of exclusion and a boundary marker; language competence alone is not enough. It might be argued, however, that not giving people the opportunity to learn the language makes them unable to use Norwegian for instrumental purposes, too; they are left outside (*utenfor*) from the very beginning.

Olaf: The language is extremely important. I criticise the community (*samfunnet*) for not taking care of this. It's a societal responsibility, not a responsibility of those few companies up here. And some do something about it, and some unfortunately don't. [...] People have been living here for 20 years and they don't speak Norwegian. [...] Using also this – as some do – as an excuse, that's bullshit. We have to do something about it all of us, we can't just push problems out of sight. [...] I think social dumping in Longyearbyen is much worse today than it was five years ago. There is much more exploitation of people. And I say this is only and exclusively the authorities' fault. Because they haven't put into force regulations that would prevent this. [...] You have let people be cynical. [...] Everybody agrees it's bad. But nobody does anything about it.

In Olaf's blaming the authorities, he refers to both central and local ones. The feeling that responsibility lies with the authorities is shared in town, which has to do both with the company town legacy and with the Norwegian majority's view of the state as the ultimate guardian of the public good. While in the case of climate change adaptation, the authorities take over the responsibility (Meyer 2022), in the case of inclusion they refuse to do so.

In many ways, the invisible fences (Gullestad 2002) erected by egalitarian cultural themes tightly related to the majority nationalism and racism known in Norway started growing also in Longyearbyen once it became a messy mesh instead of a neat and controllable company town.

However, both Susan, in her reflection on the Svalbard environment as potentially inhospitable and encouraging solidarity and cooperation of its human inhabitants, and Wenche, in her fear of parallel 'bubbles' express something particular about Longyearbyen. The town is indeed exposed to a number of vulnerabilities (energy, health care, search and rescue, military tensions, climate change and more). In other words, there is enough to worry about anyway, and if barriers between people keep growing, the feeling of discontent with life in town follows the trend. Jørgensen's (2019) argument about communitification touches upon emotions that arise when a 'community' is dealing with change and which shape the fight for desired futures. When Olaf, Wenche, Simon or Susan refer to an entity that is weak/weakening, it is *fellesskap* – something to keep the community together. The political goal of sustaining the town is free of values such as care, engagement, knowledge or continuity, which Longyearbyen residents identify as important to thrive. The aspect of change is perceived as obvious and overwhelming: the town has become more diverse; the local governance has been transformed; knowledge and continuity are perceived as being lost; and the turnover is experienced as impacting more. These processes cause problems people see as negative for *fellesskapet*: the language barrier, uneven working and housing conditions, exploitation of the non-Norwegian workforce, formation of 'parallel communities' and the push for Norwegianisation.

'What Do They Want with Svalbard?' Heading Towards an Uncertain Future

> The boundaries as well as the values, meanings, and ideas of communities dynamically appear, change, and disintegrate, and [...] these transitional movements are dependent on the emotions associated with the given community at the given point in time. [...] In order for the community to be able to draw on the social capital as a resource, the community's boundaries must be continually affirmed and reaffirmed in exchanges. Active future-making is a possibility. [...] The potential benefits of such communitifications are considerable, in so far as the successful demonstration of boundaries may come to demarcate who is to be allocated what resources. [...] Emotions are key to understanding these social dynamics; they frame what *uchronotopias* can be narrated, and they may, therefore, impact heavily on the communities' abilities to create desirable futures for themselves. (Jørgensen 2019: 9)

In the concept of communitification Jørgensen developed in the context of Arctic extractivism, active future-making is presented as desirable. In my encounters with Longyearbyen residents of different nationalities and professions and of varied Svalbard age, a feeling of helplessness, loss and confusion emerged about what Svalbard's 'desired futures' are. The trouble with local community – the mismatch between the goal of sustaining a Norwegian community (meaning settlement) and the wish of Longyearbyen residents to conjure up a community laden with different values and emotions – hinders a bottom-up formulation of *uchronotopias*, 'narratives connecting place with a vision for a perfect time to come' (Thisted et al. 2021: 1).

When I met Sunniva who shared with me her imperfect Christmas tree metaphor to illustrate what she feels 'the local community' is in need of, I was new to the field and had little auto-ethnographic material to reflect upon. Almost a year later, when meeting Olaf, I had been through the struggle with the language barrier, efforts to become part of 'bubbles' that were difficult to access, and more than 100 conversations with people sharing their lived experience from a place in the midst of change.

> *Zdenka*: I hear there should be mostly Norwegians living here. I am struggling to understand what that means. Am I a threat as a non-Norwegian living in Longyearbyen?
>
> *Olaf*: I don't understand this either. Frankly, I think it's about money. It's just to harvest (*tine*) money from the big society (*storsamfunnet*). It's used as an excuse to get economic support [...]. We have become so self-centred and cynical in our community (*samfunn*) that we do whatever to reach economic goals. [...] We need a common goal. Without a common goal we won't succeed. But today we pull different strings and nothing happens.

Olaf's common goal resembles Sunniva's tree; it is a call for a narrative that would connect instead of a polarising one. Such a 'local' narrative might again risk falling into the pitfall of 'equality' – meaning rather 'imagined sameness' – a feature of the Norwegian egalitarian culture so aptly analysed by Gullestad (2002). Whether it might be possible to build up a vision for a desired future Longyearbyen inhabitants could identify with across their different opinions and agendas is unknown; there is no space to try. During one of our research conversations, Wenche came up with another metaphor touching upon the same issue.

> *Wenche*: If you would think about it as a boat.... There is an expression saying 'We are all in the same boat' (*vi er alle i samme båt*). So a boat needs a hull and some deck joints that are crucial for the boat to float. To move forward. To exist. [...] And then somebody must be the captain, somebody who has a direction in mind, a vision. And somebody must push the boat forward, you need a propeller, an engine, a fuel, well something physically and economically must be in place. [...] And there is space for quite a few on board. But just a few are the officers and more can be the crew. And some are just guests and passengers. There is space for many, but people have different roles. And I think there must be at least some scaffolding, a certain framework for the leadership to navigate the boat. You need a course and a map for the navigation, and a clear delineation of roles and responsibilities. [...] And I think that's where Longyearbyen is struggling. A clarification in roles and expectations is missing.

In the wish to pursue a vision for the future of the 'community', both Olaf and Wenche recall the central authorities who are seen as co-responsible for the lack of a clear direction and for a blurred legal landscape that allows negative patterns to evolve while singular entities (individuals or companies) profit economically. When I asked Simon about his potential agency in a situation that he describes as discriminating, he held back:

Simon: As a minority, you always have this disadvantage. [...] It's just the reality that you have to live with. [...] You just have to accept it.

Zdenka: But can you do anything about it? I mean here. Is it possible to play an active role in all this?

Simon: Personally? I don't think so. [T]his is a big political-level game play, me as a single pawn in the game has nothing to do with it.

When listening to my interlocutor's reflections, I reminded myself of Hanne's scepticism about social scientists fishing out sensationalist notions of fast change, just like journalists. But I could not help myself sensing that the climate is indeed changing. Arne Holm, political analyst and editor-in-chief of *High North News*, wrote a commentary after the Ministry of Justice approved the suggestion to take away voting rights from non-Norwegian residents of Longyearbyen who have not lived on mainland Norway for at least three years, with the title 'While Temperatures Rise, Relations Between People Grow Cold' (Holm 2022).

What is more, the increased mobility determining life in Longyearbyen is more 'likely to be informed by a general ethos of movement and personal development that emphasises social disjunction than one which emphasises continuity' (Amit and Rapport 2002: 35). Social actors who gather in groups, personal networks, subcultures, 'bubbles' and 'parallel communities' of Longyearbyen 'can call into play cultural imaginings: categorical identities, notions of home, belonging or community' (Amit and Rapport 2002: 23). Mobility and transience, typical features of Longyearbyen exacerbated since the turn of the millennium alongside globalisation, dampen the capacities of Longyearbyen residents to mobilise politically.

While this poses a major threat to people who wish to claim agency in shaping their home's future, it is a convenient strategy for the external governance. 'What do they want with Svalbard? Do they actually want people to live here?' people would often ask when confused by measures that, instead of mitigating disempowerment and transience, reinforce them. The logic of the state, though, stops at the point where the condition for claiming sovereignty through Norwegian presence is fulfilled, and is uninterested in the patterns of the population's texture in the neighbourhood. But what the government likely sees as a nuisance, a noise in the background of the fanfare celebrating the green turn of 'the miniature of Norway', the anthropologist feels the urgency to document and think with.

8

In the Neighbourhood

Well the eggs chase the bacon
Round the fryin' pan
And the whinin' dog pidgeons
By the steeple bell rope
And the dogs tipped the garbage pails
Over last night
And there's always construction work
Bothering you
In the neighborhood

—Tom Waits, 'In the Neighborhood'[1]

I am sitting in *Kulturhuset*, one of the many venues where cultural events happen throughout the year. Longyearbyen is very rich in terms of culture and sports. *Dugnad*, the Norwegian institution of volunteering for free-time initiatives is one of the key pillars of community life. In my interviews, people often mentioned that the *dugnad* spirit has decreased since the town became so international. Newcomers sometimes strive to fulfil the expectations of the culture of the majority, and dedicate free time after work to activities such as baking cakes, fundraising for charity, volunteering as sport trainers and helpers of all sorts when events for children are being organised. But some do not, which puts more pressure on those who do, and contributes to the feeling the town is no longer how it used to be.

An important part of Longyearbyen's cultural life is the music scene. There are four choirs, an impressive number for a village of 2,500 inhabitants. The town also proudly hosts several bands, including a big band orchestra, and participation of local amateur ensembles is traditional during major music festivals of international reputation. Music events often mark the end and the beginning of the dark season, with jazz and classical music in February and blues in October.

---

1. Tom Waits, 'In the Neighborhood' © 1982 Jalma Music.

It was late January 2020 and I attended the famous *vorspiel* of the Polarjazz festival. A small ensemble of local artists performed 'In the Neighborhood' by Tom Waits as one of the final acts, and the audience, by a vast majority Norwegian, seemed thrilled. The lyrics fitted, in an amusing way, with Longyearbyen's lived reality, with a lot of noise due to construction work, shabby surroundings, trucks and guns, empty shelves in the store, Filipinas in the church, and badly maintained houses. When I was leaving the venue, one of the performers, who also is a local politician, addressed a comment to myself and the colleague I was chatting with, another social anthropologist on fieldwork: 'You guys who are supposed to write about us, you must mention how unique this place is! So much culture in such a small and remote neighbourhood!' On my way home, I walked together with a friend, a non-Norwegian freelancer who had been living in town for several years. We both noticed the same phenomenon – the tension between the extraordinarily rich cultural life cherishing the diversity of the town, and the striking absence of some segments of the population at events like the *vorspiel*. My friend shared some ironic comments regarding the 'Neighborhood' song, as she was mostly acquainted with people who struggle with inclusion, or who feel marginalised or directly discriminated. I was negotiating with myself: What role does the gap between 'Norwegianness' and 'non-Norwegianness' play in the life of the town? When we stopped to say good-bye, my friend shared an anecdote. She had been at an informal gathering and the presence of the anthropologist (myself) popped up in the chat. My friend was asked by a friend of hers: 'Do you think [the anthropologist] will surrender?' The dialogue continued as follows: 'Surrender? To whom?' 'To the Norwegians.' Despite that night's biting frost, we engaged in a long conversation about what this question actually meant and what the connotations are of the few words uttered in this short exchange. Why the war-associated verb 'to surrender'? And who are 'the Norwegians'?

I went to bed, cold and puzzled, trying to make sense out of the neighbourhood where people of many nationalities claim they are deprived of agency, making their lives unliveable, and where growing segregation makes one draw thick imagined lines, Gullestad's invisible fences, dividing insiders from outsiders. How do I write about Longyearbyen? Whose representation do I pick, the one of the enthusiastic musician, or the one of my frustrated friend, tired of having to prove her worthiness through partaking in *fellesskapet* or learning Norwegian? None and both, I thought and fell asleep.

The discourse of Norwegianness has 'hard' and 'soft' versions in Long-yearbyen. In the rhetoric of the government, the main goal is to have Norwegian passport holders coming and going as the vast majority of the population, to claim presence and thus foster sovereignty. To achieve this goal, the work and housing markets are regulated in ways that barricade them off from non-Norwegian residents, who are marginalised through racialised and extractive jobs, restricted access to health, social and admin-istrative services, and limited and expensive housing. During the two and a half years of my fieldwork, there was a significant pattern in my inter-views showing that people are more and more concerned with what kind of rhetoric is being accepted in the public discourse:

> You have Norwegians saying that there should be more Norwegians here. And that there's too many foreigners here. You can describe that however you want but in my eyes this is pretty extreme and pretty narrow minded. People would call this racist.

Locally, the fight for Norwegianness also translates into the dominance of language and cultural values.

When I immersed myself in the social life of the town and started to find ways to read the dynamic and complex structure of the population, I became interested in one of the changes that came with the shift to tourism and that smoothed global mobility: the establishment of Asian migrants. As Henrik, my acquaintance from the mining community, for-mulated it:

> you can also have that point of view saying that only Norwegians can be part of the local community here. But I think that's unfair. Those people are here because they are needed here. They are doing jobs that Norwegians don't want to do. I don't think they are stealing jobs from Norwegians, I think they are making it possible for Norwegians to be here.

Sunny's Life in the Neighbourhood

Sunny was the first person from Thailand I interviewed in Longyearbyen. I saw her posts on a Facebook page where locals sell and buy all sorts of goods. I thought she was leaving as she was selling household items, so I contacted her. Sunny replied she was only moving from one apartment to another, but asked what the interview would be used for. When I explained the aims of my research, she replied: 'If it will be good for some people I am willing to do it. I like to meet new people and new attitudes, too, to keep myself motivated.' She helped me get in touch with other Thai residents of Longyearbyen. It was thanks to Sunny that I got my first insights in the life of Thais in Longyearbyen.

Sunny had been living in Svalbard for a few years. She came alone, after a Filipina she was acquainted with told her 'Why don't you come, there is no visa needed, no work permit, it's just cold.' She had a Bachelor's degree in Agricultural Economics, but she started to work as a cleaner. She switched jobs every few months, all in the service industry relying on tourism. When I asked about her motivation to move to Svalbard, she said the primary one was money and her family, whom she was able to support with remittances. But after staying longer she started to feel attached to the place, which she described as very peaceful, reminding her of her life back in rural Thailand where everything is reachable on foot.

But despite some benefits, life was not easy. Many new migrants from Asia, mostly Thailand and the Philippines, had been coming to Longyearbyen since she arrived, and they were struggling with finding a job and a place to stay. Sunny was convinced the segregation many experience was caused by the absence of a common language – a public Norwegian course was abolished a few years ago and only a few Thais spoke decent English. She was concerned and even angry:

> We feel like inside we are lower, we are from the Third World, in our country there is inequality. The discrimination is in our head, it is scary for us to lose face, if you do something wrong in the public, that is the worst thing that can happen to you if you only stay with Thai people.

By the time I was meeting Sunny regularly, I was randomly approached at a conference held in Longyearbyen by two women in their twenties who introduced themselves as Planning Officers of the Tourism Authority of Thailand, an organisation under the Thai Ministry of Tourism and Sports. Their visit to Svalbard, they told me, was motivated by search for inspiration on how to promote Thailand's tourism industry. We went out for a drink and had a conversation about the issue of 'service mind', as they used the term. When I encouraged them to elaborate on the meaning, they informed me it was the key for the success of Thai tourism. People, especially wealthy tourists from the rich global North, come back to Thailand not because of the beautiful nature (which can be found also elsewhere in South East Asia, and perhaps even more stunning) but because they love the 'service mind' of Thai tourism workers. 'We are taught to treat our guests as gods.' Blue eyes and blond hair are believed to be indicators of wealth. My conversation partners from Bangkok wondered whether this belongs to the key factors in the success story of the Thai population establishing itself in Svalbard.

It might play a role. In any case, Sunny was unhappy about the 'service mind' mentality being imposed onto her. She wanted to break free from it and find a job where it would not haunt her, but she hit the language barrier. She had no chance to learn Norwegian through her employer. She understood that if she could speak the language her chances of succeeding in the overheated job market would be higher, and felt frustrated when she realised her only option was to try to learn the language by herself. She came without a network or family back-up in the town, which is rather atypical in Longyearbyen, and soon realised how disadvantaged she was by that. She felt discouraged by emails potential employers never replied to, and slowly lost trust in the presumed fairness and transparency of how life works in Longyearbyen.

It was not a dream to be in Longyearbyen, but it was not a dream to return to Thailand either – back to the corruption, low salaries and miserable job opportunities. In Svalbard it was cold, there were no trees, no hanging out while eating street noodles, and a lot of fatigue and loneliness. 'We take this suffering just to have a good life in Thailand. People complain Thai people don't spend money, but they have to take care of their families.'

Working in Longyearbyen was also to the detriment of Sunny's private life. 'I don't have the heart to find a man, I have too much responsibility back home.' Her mother was seriously ill and her sister was looking after their mother, so Sunny also had to provide for the sister who could not work, and her two children. She also mentioned 'It is painful to think about my nieces that now are forgetting me.'

The burden grew as, after some time, she was also helping one of her relatives establish themselves in Svalbard. Sunny was trying hard to build up a business so it would be easier to support so many family members.

I would like to see everybody happy, that is why I am struggling here. [...] We should have the right to say what we need, we only work and go home and sleep, work, go home and sleep – but I want more from life!

[...] My dream is to be the real voice of the Thai people. Not the voice only for my own benefit. What is the point of being a voice if you only get everything for yourself?

Sunny suffered from political disempowerment, public invisibility, and the assumption of her readiness to accept inconvenient working and living conditions. 'Many Thai women find a Norwegian partner for convenience, not for love. And then they have to do everything they are told to. Do you understand?' I thought I did. From a dream, life in Svalbard became unviable for Sunny, and she decided to return to Thailand. Since then, we have had only sporadic contact and, apart from snippets of her life I get a glimpse of through Facebook, and a short email exchange regarding approval of my written framing of her story, we have not kept in touch.

Migration to Svalbard from Thailand and the Philippines

Even though mentioning that the biggest minorities in Longyearbyen consist of Thai and Filipinx nationals is usually met with astonishment, migration from South East Asia is not uncommon in other Arctic places. The legal landscape created by the Svalbard Treaty is an obvious catalyst for migration. The case of Thai and Filipinx migrants to Iceland (Bissat 2013; Skaptadóttir 2010, 2019) serves as a good comparison with Long-yearbyen, with some similar patterns, but also some profound differences. As mentioned earlier, international migration to Svalbard has accel-erated and tripled since the turn of the millennium, with a significant increase of Asian migrants. I was not the first social scientist to be intrigued by the phenomenon, but the last studies with a similar focus were pub-lished more than a decade ago (Moxnes 2008; Jensen 2009). Then, there were almost no migrants from the Philippines in Longyearbyen and the situation was in some key aspects different from today.

The clash between the geopolitically motivated narrative about Long-yearbyen, and the lived experience of Thai and Filipinx migrants is decisive in understanding their social participation (which is often lamented as too low), what mechanisms hinder inclusion, and why earlier inclusion measures had been cut off. While these migrants constitute a workforce required by the service industry, where they can hardly be replaced by Norwegian nationals, their lives contest the imaginary of Longyearbyen set out by the Norwegian government. When I asked Simon, a scientist encountered in the previous chapter, about his relation to Asian migrants, he exclaimed:

> I don't think the way they are treated is fine. But it's Svalbard, which means everybody is welcome to come here. And sure, Svalbard is governed by Norway and it is part of the Kingdom, but it's not Norwegian mainland. Otherwise we would have had our rights. I actually think all these migrants are very good for Svalbard because they offer some sort of resistance to ... yeah, a gentrification of the entire area. Just filling it up with Norwegians. I think they are the best stronghold that Svalbard has to prevent Norwegians from a hostile takeover, replacing the whole population with Norwegians.

In Simon's straightforward view, international migrants are those who disrupt the idea that only Norwegian polar adventurers and nature lovers settle in Svalbard. While Norway tries to encourage the influx of Norwegian migrants, some non-Norwegians stay over long periods of time despite the hurdles they meet.

To be able to theorise the lives of Thais and Filipinx in Longyearbyen, how they perceive the changing place and what kind of perceived change their presence induces, I needed to do ethnography with them. I paid Sunny for the time she spent on facilitating meetings and translating during interviews. It was the everyday experience I was interested in, and I asked mundane questions – how people lived, what brought them to Svalbard and what kept them there, what kind of difficulties they were facing and what strategies they used to tackle those, how they perceived their living and working conditions, to what extent and how they kept up ties with the country from which they emigrated, how their use of language(s) evolved with time spent in Longyearbyen, how they practised religion, and spent their free time.

The language, in fact, seemed to be a very big thing. Before I continue with the story of Thai and Filipinx migration, an intermezzo with the language of power is necessary.

'Snakk norsk!' *Intermezzo with Language as a Boundary Marker*

After a few months in the field, when I still spoke broken Norwegian but had improved my understanding, some of my interlocutors preferred to answer in Norwegian to my questions formulated in English. I felt it was fair enough as long as both conversation partners agreed this was done for mutual benefit – them not having to struggle with English vocabulary and able to express themselves freely in the language they preferred, and me improving my language competence through being exposed to Norwegian upfront. In rare but unforgettable cases I felt the relationship was uneven and the interview situation awkward – I was being lectured through the use of the language of power about 'to whom this place belongs'. This was an experience I had in common with some of my Thai and Filipinx interlocutors; the feeling of intimidation and inhibition associated with a missing or imperfect language competence.

As Olaf shared in Chapter 7, and several others in similar words, people lived in Longyearbyen for ages without learning Norwegian. The issue of people with migratory background integrating with the majority through language is a heavily studied topic in anthropology, migration studies and other disciplines, within the Nordic context as well. What language is, what it does with human communication, thinking and world making is one of the foci of linguistic anthropology. The issue of language was not my research interest, but it soon became a major recurring theme in my interviews and a working tool I could not do without.

Before we translocated to Svalbard, we talked to a few acquaintances about practicalities, among other things the language course, which at that time did exist. When the date of our moving became clear, I contacted LL to ask about services such as a kindergarten and a language course. While my question about child care was answered and our application processed in an accommodating way, I was informed the language course had been abolished. This took us by surprise, but we did not dwell on the lack of service too much, busy with other errands to run before we move.

It was only when the fieldwork started that I realised how resentful some people felt about losing the language course offer; some, for example Asian parents of children who attended the local school, were desperate about not having any opportunity to improve their knowledge of Norwegian. As it became an issue for my own work, too, together with a few other non-Norwegian women, we started to meet regularly with Kristin Furu Grøtting, a health worker who volunteered to help us with pronunciation, grammar and insights into Norwegian history and culture. It was difficult to find a space where meetings could be held; first we went to each other's places, then Kristin arranged we could gather in the hospital's meeting room, until we finally found space in the public library. There was a Filipina and a Thai woman attending the do-it-yourself course, and even *Svalbardposten* reported about the initiative (Solberg 2020). More people asked if they could join, but the format did not allow for growth. After a few weeks, I found myself using much of my time trying to figure out what to do. When I posted on the local Facebook page asking people to express interest in launching more such groups, the response was over-whelming; within a few days, over 60 people said they were in need of a language course, and if a regular paid-for one was not available, they would be happy to participate in a free one based on volunteering. Three kindergarten teachers agreed to help us, and in August 2019, we had an information meeting in the kindergarten. There was a handful of people in the classroom who seemed not to understand the issues discussed. Most of the people who came were Thai, and a significant number of them did not speak English. The translator could not show up due to a work shift she had to attend to, and I found myself in a tricky situation in a room with about 20 people wondering what I had to say, without a common language to share the message. We improvised, formed two groups, each of about eight people (some of the attendees had an intermediate knowl-edge and starting with the basics was not what they were looking for), and agreed when the first meeting of each one would happen. Both groups continued to meet for a few months, before the kindergarten admin-istrator responsible informed me that LL does not allow the activity to continue; when I asked about the reason, the person did not know exactly. In every case, there was no space there any more. This was also the time when the teachers reported that fewer and fewer people came to the actual meetings, and we could not figure out whether it was because of work shifts they had to give precedence to, lack of motivation, or both of these, or something else. Another teacher offered to host the course in her kitchen, but that did not seem like a sustainable solution either.

On the one hand, there was a clear need of a language course. On the other hand, a bottom-up initiative was being stopped. Later that autumn, there was a Norwegian language teacher, thus a professional instead of volunteers, who was willing to open a paid-for course in Longyearbyen. 'We are saved,' I thought. But I was wrong. When Ellen Solbakken started to inquire whether she could possibly rent a room for a reasonable price from LL, her request was declined. We were utterly puzzled. These events unfolded around the time of local elections, and several of the politi-cians with whom we met to talk about the issue agreed there was a public interest in substituting the language course earlier offered by LL with something commercial, yet affordable. Nevertheless, Ellen could not find any space for a long time, while she already had several groups at various levels of language competence ready to start. It was through personal connections ('you have to know the right people') that she found an office to rent; not very cheap, but there was no other choice. Our group with Kristin went on as before, the two groups in kindergarten faded out, and a paid-for language course began.

When the pandemic hit and tourism ceased completely in Svalbard, it was largely the group of Ellen's clients (with the exception of the scientists) who were facing major insecurities in terms of work and income. Renting the office over the summer, without any clear prospects regarding how many of the fully or partly laid-off people would sign up for a course in the autumn, Ellen gave up and left Longyearbyen. Those who were highly motivated switched to online solutions, which was indeed what the pandemic stimulated. People like those who came to my clumsy meeting in the kindergarten were left without any options.

I was acquainted with a guide, Roby, one of those living on the island permanently. When he arrived in Svalbard, he immediately started to learn Norwegian and quickly achieved an advanced level. Roby was full of ideas, a creative mind passionate about his new home, optimistic about how he could contribute to 'the community'. He told me about a party at a friend's place where he arrived, tired from work. While he would normally prefer to speak Norwegian in public settings – respectful and loyal to the majority as he explained – that day he felt too exhausted and happily started to chat with a friend in English. 'Snakk norsk!', he suddenly heard from somebody else in the room who had noticed there was a conversation unfolding in English. It was the moment Roby realised his efforts would never be enough; the language of power was penetrating informal spaces as well, and you were expected to behave all the time.

'Of Course We Migrate': Lives in Spite of Big Politics

During one of the meetings with an LL employee, Vera recollected a meeting that ARTICA Svalbard, a local art residency centre, arranged between bureaucrats and Asian migrants in a restaurant. The aim was to facilitate an arena where people who normally live in what some call 'parallel worlds' could talk. The shared perception among public employees administering tasks such as child care or health care was that moving to Svalbard from Thailand or the Philippines made no sense. Why struggle in such a hostile place, remote and cold, with amputated social services, if you are not interested in *friluftsliv* (the Norwegian concept of 'outdoor life'), cannot understand the language, have an unstable job and cramped housing conditions? But when the invitees to ARTICA's event shared their thoughts and explained where they migrated from and the ways they see their life prospects as brighter when living in Longyearbyen, Vera understood: 'Of course they migrate.'

What intrigued me was the paradoxical layers of the picture. On one side there is the rational decision to migrate and find ways in Longyearbyen, which appears as 'natural' when contextualised with the lived experience from the country of origin. Transnational ties, such as improving the life of family members in home countries through sending remittances, investment in real estate or businesses there, but also securing more opportunities in the lives of the children are among the strongest drivers of migration. On the other side, there is the politically motivated process of cutting and not introducing measures to foster inclusion, comprehensible only against the backdrop of the geopolitical interests of Norway in Svalbard. I use the term 'inclusion' here as understood by Uusiautti and Yeasmin (2019) or Karlsen (2021). Inclusion is 'an essential ingredient of overall community well-being and resilience' (Uusiautti and Yeasmin 2019: 5), having both formal and informal aspects. It signifies participation in social and political life, and access to available opportunities, services and resources. Focusing on inclusion means studying 'the conditionality of legal status and social and institutional processes of boundary making' (Karlsen 2021: 4) in a society where migrants are in a precarious position. The situation of South East Asian migrants in Longyearbyen is to a large extent similar to the situation of international guides discussed in Chapter 6. Since most of them move directly to Svalbard and have never lived on mainland Norway, they only get a so-called D-number (in Norway a temporary solution for registration) and their access to health care and social security is dependent on their employers in Svalbard. Not having or losing a job means limited (after some time no) access to these services. According to existing legislation, moving back to the country of origin is recommended in such cases.

To document the lifeworlds of Thai and Filipinx residents of Longyearbyen felt important, but also ethically challenging. I became aware of the risk of mistrust and suspicion thanks to Sunny, who was generous in sharing her opinions and explanations of how she understood the 'codes' and legacies of the past, and I met that worry repeatedly later with other participants. The potential danger my Thai and Filipinx participants fear has several causes. During past decades, as the numbers of migrants from Asia to Svalbard grew, several issues related to a sort of human trafficking, social dumping and exploitation were discussed publicly (Ylvisåker 2016). Local politicians responded to these issues with varied success, and not everything known about is actively tackled. In addition, as the settlement is small and many Thais and Filipinx have relatives in town, private issues among my potential participants complicated the research. Last but not least, I have been repeatedly told Thai women and Filipinas living in a relationship with Norwegian men are not willing to talk to social scientists as they do not wish to be represented as exchanging their bodies for an economic upgrade. Another obvious barrier was the language. After a few unsuccessful meetings, I occasionally asked a local translator (Sunny, or other people after she had left) for paid help.

The first citizen of the Philippines who probably arrived sometime in the 2000s to reside in Longyearbyen is not widely known, but the cleaning business of the first well-established Thai woman who came to Longyearbyen about 30 years ago, married to a Norwegian man, is still operating. But Filipinx established themselves faster thanks to their advanced English. Miners, and later workers in the construction industry and beyond, brought their Asian spouses to Longyearbyen. Apart from women who migrated to Svalbard to join their Norwegian partners, most people born in Thailand or the Philippines living in Longyearbyen today have been encouraged to move to Svalbard by close or distant relatives. Yet some came because of a friend's idea, or even only information online about Svalbard being a visa- and work-permit-free area.

As Bissat (2013: 52) summarises:

a few Thai women with Norwegian husbands anchored the stream in the 1990s, and then recruited both male and female workers from Thailand in roughly equal numbers. Thus, the current gender ratio among Thais in Svalbard is reportedly more balanced than that of countries where marriage (by women) is the only viable form of access.

Though I have objections to using water-associated terms such as 'waves' or 'stream' when discussing migration, the core of the finding is valid also for Filipinx in Longyearbyen. There are quite a few Thai-Norwegian and Filipina-Norwegian couples (no statistics available), and multi-ethnic relationships have been part of life in Longyearbyen for decades. Mistreatment as a result of a trade-off partnership has been reported recently (Malmo 2021).

Among my participants, moving to Longyearbyen to find a job at a place with a family network was the most common motivation. Finding or joining a Norwegian spouse is hardly ever the primary driver, but it can of course happen that relationships develop. A typical thought process is comparing the possibilities in Svalbard with one's chances of living a dignified life in a country where it is hard to find a stable and well-paid job, where a good education and standard health care requires large financial investments, or where life is not safe because of high levels of criminality. 'We call it "the Golden Opportunity",' a Filipino friend shared. The diligence and determination of East Asian migrants is appreciated by employers, and the hard work and dedication is motivated by making a living not only for oneself, but for others one cares about in Svalbard or elsewhere. As Skaptadóttir (2019) shows, it is also a typical pattern in other Arctic locales with transnational migrants: 'The ability to care for those left behind, be it children, parents or others, is often an important reason for working abroad. Sending remittances can, in addition, boost people's social status in the country of origin' (2019: 213).

There is one more specific segment of Thai and Filipinx nationals who did not choose to translocate: children. As of 6 April 2021, there were 11 Thai residents younger than 18, 8 of whom were born in Norway, and 28 underage Filipinx, 4 of them born in Norway (Tax Office, personal email communication). While the numbers of Thai children have lately been relatively stable, there has been a major influx since 2015 of Filipinx children who joined their parents. Most of these children arrived when they were school age, some as teenagers. Both the pre-pandemic developments in tourism employment and the impact of the Covid-19 pandemic (making some people stay who would normally only be seasonal workers) contributed to higher numbers of Filipinx children.

The connection between the determination of both Thai and Filipinx parents to keep working in Longyearbyen, and their hope for a better future for their children is evident. All my participants who have children stated that the chances of providing their offspring with a less challenging life is the major driver for their resilience. The parents identified good and free education, the opportunity to learn both Norwegian and English, and gaining skills that enable their offspring to find a job in Norway as the main reasons for considering Longyearbyen a good place for their children.

People in the first generation travel to the country of origin to visit their family at least once a year – mostly during the dark season, which is also the low season for tourism in Svalbard. They often look forward to the day they might finally return to the place they have been missing while in Svalbard. But it is equally common to hear that the plan is to stay for 10 or, in case of young parents, even 20 more years. Plans to stay in Long-yearbyen for a decade or more were much more common among my Thai and Filipinx participants than among Norwegian or other European participants. The typical milestone would be when the child or children finish their studies and can live on their own. Fewer parents are hoping for a future for their children in Longyearbyen, more for a future in Norway. Hardly ever did parents say they wanted their children to return to Thailand or the Philippines. Specific is the situation of Thai women and Filipinas married to Norwegian men; these multi-ethnic families divide their time between Norway and South East Asia, able to live 'a good life' in both countries, to which the children born or raised in these relationships also keep a bond. This practice of the spatial splitting of family life was temporarily interrupted by strict travel regulations in 2020 and 2021.

When the town became paralysed by severe restrictions at the outbreak of the pandemic in spring 2020, and local politicians worked hard to get governmental funding for so-called 'third-country citizens' to get flights 'back home', interest in the scheme was much lower than expected. The local authority spent NOK 2 million of the allocated NOK 7 million on tickets. In total, 20 people were granted funding for a one-way air ticket, most from Thailand (Bårdseth 2020). Several of the 20 applicants returned to Svalbard after just a few months. The administration soon noted the factor of having children in Longyearbyen was decisive for non-European residents to reject the offer, because if an adult applied for the grant the whole nuclear family had to leave (Bårdseth 2021a). As one of my interlocutors explained, 'people don't want to go back because there is no future for them. Most of those who travel back are single or have no kids. Or they have children, but they are grown up and can survive on their own.'

Those who did not succeed in establishing themselves in Longyearbyen well enough and left their children in the country of origin to be brought up by the children's aunts, uncles or grandparents, still claim the years spent in Longyearbyen – working hard to be able to support their children's studies – were worth the emotionally exhausting effort. Research interviews where my participants shared with me their reasoning for these decisions were demanding for me, and made me realise how grateful I was to live in Longyearbyen with all the three boys by my side.

It is more common among Thai women and Filipinas to have a university degree than among their male co-citizens. Just as in the case of the

Filipinas in Iceland discussed in Skaptadóttir (2019), it is difficult to find a job in Longyearbyen where they can use the expertise gained through education in the countries from which they migrated. The reasons are the language barrier and Norway's strict rules around acknowledging diplomas from 'third countries'. As Gullestad (2002) shows, this is a known phenomenon in mainland Norway:

> Many 'non-Western immigrants' work in unskilled and semiskilled occupations as taxi-drivers, hotel personnel, cleaners, and so on, doing many of the jobs that 'Norwegians no longer want'. Educated 'immigrants' often experience difficulties in obtaining employment that fits their educational level. (2002: 47)

Despite being an interior designer, an agronomist or an accountant, Thai and Filipinx in Longyearbyen are limited to getting jobs in the spheres of cleaning, shop assistant, catering (hotels, restaurants and cafés), massage studios or the Thai store. To establish a competing business in the niches already occupied by those who arrived earlier is a daunting challenge. A general pattern appears, with a clear link between arriving in the locale earlier meaning higher chances of finding permanent and full-time jobs, and saving enough money to buy property (a key factor as housing is very scarce and expensive). Thai and Filipinx migrants express willingness to re-qualify and take further education, be it in Svalbard, in Norway or online, to upgrade to wider job opportunities.

My Thai and Filipinx participants, obviously with the exception of those who live in multi-ethnic relationships, mostly found it difficult to relate to the rest of Longyearbyen's population other than professionally. Many are used to living their private lives hidden from the gaze and vibe of the majority, with free time dedicated to family living in Longyearbyen, digitally cultivating relations with people in the country of origin, socialising during dining events or just resting in solitude after exhausting work shifts. Religion (mostly Buddhism in the case of Thais, Catholicism in the case of Filipinx) is practised in privacy, but some Filipinas and their children visit the local Protestant church every now and then, especially but not exclusively when there is a Catholic mass. Few of my non-Asian participants have a closer relationship with somebody of Asian origins, but this changes with age; as stated earlier, there are about 40 children living in town from these two countries, plus there are children who have Norwegian citizenship and one Thai or Filipinx parent (no statistics available) – to my knowledge in all cases the mother.

The Thai segment of the population is often seen as homogeneous from the outside by non-Thai residents. My experience suggests there certainly exist firm family bonds that substitute for the absent safety net of social welfare, but to describe the Thais as a united sub-community would be misleading. Power structures have continued to develop and a grey zone of rules and the existence of mechanisms illegible to the majority has long been known (Jensen and Moxnes 2008). Those who are well established, have a long personal history in town, have a stable job and/or master Norwegian are more competitive and therefore can act from a position of power over newcomers, especially those who arrive in Svalbard without any extended family members already on the island. 'There are more people willing to work so it's more difficult to find a job,' a Thai woman shared.

The Filipinx segment is linguistically more diverse, but most people understand either Tagalog, Cebuano or Kapampangan. Nevertheless there is a certain language barrier within the grouping, which makes life harder, especially for older children who move to Longyearbyen and have to rely on English when communicating with Filipinx peers. Many Filipinx speak English at an intermediate or even advanced level, while among Thai residents fluent English is not a typical competence. There are good reasons to believe this difference creates a competitive advantage locally for Filipinx migrants, and might partially explain their quick establishment and population growth, despite their later arrival. Another significant factor might be the long-term strategy of the government of the Philippines encouraging people to emigrate, adapt and support their network in the Philippines through remittances (Skaptadóttir 2019).

While the first generation continues to communicate in the mother tongue among themselves, there is evidence that the offspring lose language competence in the mother tongue rapidly unless parents or guardians dedicate much free time to teaching children how to read and write. Given the different script and complex grammar, the phenomenon is even more pronounced among Thai children.

Shrinking Margins of Inclusion

In a reader's letter to the local newspaper, the leader of the local branch of the national Conservative Party (Høyre) writes:

To do local politics can be challenging. One of the things I think is most demanding is that you are almost expected to have racist attitudes. [...] I get truly sad when my colleagues and friends tell me that they feel like second-class citizens in our town, because they are not Norwegian citizens. [...] This is where they want to live. [...] In Longyearbyen, it has become ok to discriminate on the basis of nationality. Foreign citizens are kept out of the housing market and they are told that they do not need to apply for jobs because only Norwegians will be considered. (Johannesen 2019, translation mine)

In the writings of Pedersen (2017, 2021), internationals are portrayed as performing activities that threaten Norwegian sovereignty over Svalbard. Such a representation has major implications for inclusion. As already explained, non-Norwegian population growth accelerated as a result of a combination of factors. Opening up locally for more research and tourism, and global developments in communication, lifestyle and transportation turned Longyearbyen into an attractive and accessible place for people worldwide looking for jobs, adventure, life in the Arctic, a new start, a better future for their children and beyond. In 2007, the town was awarded the prize 'International Community' by Norwegian Crown Princess Mette Marit, which today is either forgotten because of the high turnover, or bitterly remembered as a different era, when it still was normal to appreciate the diversity of the town's population. Since then the numbers of non-Norwegians, especially Thais and Filipinx, grew sharply and a strength turned into a supposed weakness. The result of these contradictory processes is deepening segregation and a deteriorating social climate.

In the comments shared by my Thai and Filipinx participants, three issues were perceived as most urgent. First, as Norwegian legislation regulating working conditions is only partially valid in Svalbard, people without permanent full-time jobs often accept conditions that are immoral, and they are exposed to vulnerabilities related to the limited housing market. My participants would share with me accounts of working hours that had never been paid for by the employer, or wages of Thai employees working at the same place over 10 years being equal to the wages of newly arrived Norwegian employees. People active in the trade union that has a local branch in Longyearbyen claim it is on the verge of impossible to recruit Thai and Filipinx workers to join, and they ascribe it to cultural differences. When the pandemic resulted in hundreds of people being laid off, many Thais and Filipinx were forced to give up their rented flats and adjust to even more cramped living conditions. It is perhaps here that it might be a significant comparative advantage to find a Norwegian spouse with housing granted through his employer.

Second, the language barrier is growing as the government states that the local authority is not obliged to offer any public course (either free or paid-for) in the Norwegian language for adults. To learn Norwegian is harder in Longyearbyen compared to mainland Norway, where NGOs, libraries and other actors offer courses. On the practical level it means no state money will be allocated to such an offer in Longyearbyen, so the course would require reallocating funds from elsewhere in the town's budget. My participants also often confirmed that when the course was available they did not prioritise it if it coincided with working hours. Speaking Norwegian upgrades one's job outlook and access to information (and thus also inclusion and informal power; see Skaptadóttir 2010). On the other hand, if the job and the money earned is the main and sometimes only reason for being in Longyearbyen, the risk of losing part of it, or even risking one's reputation as a diligent worker always at the employer's disposal seems too high compared to poorer Norwegian. There is a gap between a clear understanding of the role the language plays in inclusion across the population segments, and the low probability of tackling the issue as it would require a motivation campaign with incentives for migrants who would attend a course instead of working. While on mainland Norway there are both state organisations and NGOs doing the job, the only local offer as of today is a Language Café organised by Norsk Folkehjelp Svalbard every second week for two hours in the public library. It is mostly attended by European and American migrants who are often students, researchers, entrepreneurs or freelancers. An offer adjusted to the needs of Asian migrants with specific needs regarding learning Norwegian pronunciation, for example, is missing.

Finally, the Thai and Filipinx residents mostly prefer to go under the radar, which is a convenient strategy as long as life is unproblematic. Should issues occur, such as domestic violence, discrimination, bullying, exploitation, problems with visas (e.g. when travelling to Norway, or to Asia and back to Svalbard through Norway), or even just lack of information and confusion, it is difficult to get assistance, as became evident in a recent case of domestic abuse (Bårdseth 2021b). Even though the Governor's Office does provide help to people exposed to domestic violence, for example, the perception among migrants is that there is no easily accessible agency for advice or counselling: 'If we're in trouble we don't know where to look for help.' A Norwegian friend shared with me an experience with a Thai woman who approached her in a corridor with a piece of paper, where it was written HELP ME. With the use of Google Translate they managed to communicate what the issue was, but my friend could not act on it. Such situations might build unofficial, informal structures of mutual help, but they also allow for feelings of frustration and hopelessness on all sides, unhealthy dependencies and ambiguous power relations.

The lives of those who feel 'at the bottom of the food chain', be it tour guides, East Asian migrants, or others whose engagement with Svalbard goes beyond Norwegian work tourism (*arbeidsturisme*), which is the only form of temporary inhabitation the government supports and encourages, can be told as stories of resistance. The question of political mobilisation and partaking in the decision-making process in Longyearbyen was a lingering one when my fieldwork started. When my stay was over, Longyearbyen was buried under a heap of suggested changes in regulation, and the aim to 'increase the Norwegian ratio' has turned from something one hears about on the grapevine to a fully public strategy.

All recent changes in the legal landscape point in the direction of further exclusion in the neighbourhood: access to bankID (personal electronic identification necessary for a range of online services in Norway) has been restricted since early 2019; the only local bank was closed in late 2020 resulting in non-Norwegians moving to the island having difficulties in opening a bank account; the Ministry of Justice approved taking away voting rights from anybody who has not lived on mainland Norway for a minimum of three years (Norwegian Ministry of Justice and Public Security 2021). The hope is to dampen down migration from outside Norway; the result is a grey zone of informal power structures, dependencies and inequalities cemented over time.

*   *   *

In the final chapter of Part III, I discuss how the politics of difference, denying Longyearbyen's complexity, and fake democracy reconnect to the trouble of climate change.

9

'Make Longyearbyen Norwegian Again': Denying Superdiversity

When the concept of superdiversity was introduced (Vertovec 2007), there were almost 180 nationalities represented in the population of London of 7.5 million. In 2021, 58 nationalities were represented in the population of Longyearbyen of 2500. In other words, Longyearbyen is a superdiverse settlement. This does not simply mean 'a lot of diversity'; as Vertovec (2023) insists, superdiversity is a helpful tool when we need to talk about 'multidimensionality or intersectionality with regard to new patterns of migration' (2023: 6). In Chapter 8, I discussed the variety of factors that impact the Thai and Filipinx population of Longyearbyen; what matters is age, gender, education, religion, language competence, occupation, housing situation, type of work contract, ethnicity, but also the timing of their settling in Longyearbyen, networks of family and friends, relations with the majority and connections with people in positions of power. Thus, it is an act of violence to think about the Thais and Filipinx of Longyearbyen as united groups. This is equally valid for other migrants in town; Longyearbyen has since the 1990s witnessed:

> increasing movements of people from more varied backgrounds represented by more differentiated categories. Not only are there more, smaller cohorts of people from a wider range of origin countries, but [there are] shifting flows of people with wide-ranging nationalities, ethnicities, languages, religions, gender balances, age ratios, human capital, transnational practices and, especially, migration channels and legal statuses. [There are also] shifting combinations of these backgrounds and categories. (Vertovec 2023: 6)

The benefit of thinking with Vertovec about superdiverse Longyearbyen lies in the effect of juxtaposing the observed diversification of difference and the increasing top-down push for a return to an imagined heterogeneity of the settlement. There is a dynamic in the growing complexity of 'the neighbourhood' that is fought by the Norwegian government with the help of openly nationalistic rhetoric and slow bureaucratic violence that makes existing margins of inclusion shrink if not disappear. The narrative of Longyearbyen being 'a somewhat normal and well regulated Norwegian local community and not a sort of an international village' (Pedersen 2016: 6, translation mine) leans onto the assumption shared by mostly Norwegian political scientists that Longyearbyen's superdiversity is a threat to Norway's sovereignty claims. The effect of this narrative is an extreme simplification of the multi-layered story the town has to tell; a simplification of 'us, the Norwegians' against 'them, the foreigners'.

As I have shown in Chapter 7, this phenomenon is not unique to Longyearbyen. On the contrary, it is what paradoxically makes Longyearbyen reasonably Norwegian. Nor is it unusual that the state prefers a black-and-white story to a messy one. As Scott (1998) already convincingly argued, states attempt to make societies legible for the purpose of making them easy to monitor and manipulable. To make this happen, human–nature relationships are simplified, a modernist ideology is put into practice through the power of the state, and communities are deprived of tools to mobilise and resist the violence, which is more easily done when facing any sort of crisis (Scott 1998). As I have shown in Chapter 3, this is where climate change comes in handy.

> In sum, the legibility of a society provides the capacity for large-scale social engineering, high-modernist ideology provides the desire, the authoritarian state provides the determination to act on that desire, and an incapacitated civil society provides the level social terrain on which to build (Scott 1998: 5).

The winding movement of the *ouroboros* connects climate change with globalisation, a real environmental urgency with the state's entanglement in extractivism, taking advantage of a disempowered community. To analyse processes like those unfolding in Longyearbyen, considering 'humanity as an undifferentiated whole' (Moore 2017: 595) is not helpful. Enmeshed political and economic interests that can be studied with regard to the specific time and particular place translate into 'inequality, commodification, imperialism, patriarchy, racism and much more' (Moore 2015: 597). This cannot be done from the point of departure of the Anthropocene where difference is hard to attend to. In the superdiverse Longyearbyen, only some work and some lives are valued in the state's effort to clean up Svalbard from undesired diversification; where fluid environments meet extractive economies, disempowered communities languish. As Ødegaard (2022: 12) recognises:

> in Svalbard, as 'nature' is made into a central entity, both in the demarcation of national presence and in narratives about Svalbard as an environmental showcase, certain expectations are produced [...] to align with a particular politics of presence, another kind of settler colonialism [...] a recolonization of place through environmental management and the accommodation of a particular kind of inhabitant.

Seeing it like the Norwegian state, Longyearbyen used to be inhabited by miners (the carbon utopia of the past) while now it is the environmentalist experts' turn (the post-carbon utopia of the future). First, I argue that neither the miners nor the scientists are homogeneous groups that could or should be essentialised as Norwegian upholders of the state's sovereignty; the voices of Henrik and Bjørn testify to this as do the voices of Hanne, Lena or Kim. As Vertovec (2023: 214) puts it, 'cross-cutting, partial identities characterize the social worlds of most people'.

Second, there are many more people than just Kalvig's 'hardworking pioneers of coal mining and knowledgeable researchers': Susan, Sunny, Simon, Ole, Toni, Roby, Parita, Wenche and all the others *The Paradox of Svalbard* seeks to give space to in the public debate about Svalbard. To the state, the town's superdiversity is apparently far from fitting the narrative of the Norwegian showroom of climate solutions. This is being tackled through measures that can make the dispensable (Nixon 2011) become invisible.

Located Politics of Difference and Fake Democracy at Stake

As the lived experience of my interlocutors shows, the way non-Norwegian migration to Longyearbyen is being portrayed is mostly negative. Since people's identities are intersectional, there are several forms of subordination that may impact them simultaneously. The nationalist discourse amplifies existing racism and culturalism. The concept of superdiversity sheds light on new inequalities, prejudices and patterns of segregation; take the example of people's working conditions or housing experience:

We need to consider the local development of 'a complex entanglement between identity, power and place' [. . .]; a 'located politics of difference'. This entails examining how people define their differences in relationship to uneven material and spatial conditions. (Vertovec 2023: 37)

To unpack a located politics of difference in Longyearbyen, it is necessary to critically assess simplistic views of categories such as 'Norwegians' and 'foreigners'. According to Abram's (2018) analysis of so-called Norwegian values, the person who gets to participate in the inclusive unity of an idealised egalitarian and solidary society is 'the Norwegian (member of the nation), not the inhabitant of Norway (resident or citizen)' (2018: 91). In this light, it is possible to explain the paradoxical feeling of exclusion many of Longyearbyen's non-Norwegian inhabitants have increasingly felt. Those who would like to be included and participate are disqualified by their not being members of the imagined Norwegian nation. Even if they learn Norwegian, work for a Norwegian employer and pay taxes in Svalbard, and when they conform to what is expected by the majority as an 'appropriate lifestyle' for Longyearbyen (such as taking part in volunteering, explicitly appreciating the outdoors and cabin life, showing up at public events such as the 17 May parade, etc.), from the perspective of the top-down political strategy they stay *utenfor*.

A few months after we arrived in Svalbard, we spent a weekend at a cabin about 17 kilometres from the town. The trip, with three children including a 1-year-old involved demanding logistics and physical fatigue; we had to carry five sleeping bags, food, wood, a rifle, warm clothes and sometimes also the kids several kilometres on foot across the wet tundra. It was challenging but we made it. When I came back to Longyearbyen, I shared a few images with a Norwegian acquaintance. 'Wow, you are becoming Norwegian at record speed!' my friend exclaimed. In this well-meant and encouraging affirmation that our efforts are being appreciated, Gullestad's (2002) invisible fences revealed themselves for a moment.

Social actors must consider themselves as more or less the same in order to feel of equal value. When they thus manage to establish a definition of the situation focusing on sameness, each of the parties – paradoxically – also gains confirmation of their individual value. (2002: 46)

'Immigrants' are asked to 'become Norwegian', at the same time as it is tacitly assumed that this is something they can never really achieve. (2002: 50).

Ethnic nationalism needs to be underpinned by essentialising labels. While on mainland Norway the label is 'immigrant' (*innvandrer*), in Longyearbyen the term that works as a boundary marker is 'foreigner' (*utlending*). It is used in a totalising way, eliminating other identities and applied to all non-Norwegian nationalities. The 2021 act of taking away the voting rights from all the people who have not lived on mainland Norway for at least three years is explained in the following way:

> The framework for local democracy in Longyearbyen is [...] different from municipalities on the mainland. LL shall carry out its activities within the framework of the Norwegian Svalbard policy, and manages significant funds transferred from the Norwegian mainland. LL is thus responsible for *common interests of great national value*. Participation in the activities of LL requires a *good knowledge of these prerequisites, of the Norwegian language, culture and social life in general*. (Norwegian Ministry of Justice and Public Security 2021: 5, translation mine)

During the public hearing about the suggestion, over 50 reactions were delivered to the ministry. All the reactions of Longyearbyen residents and local associations were negative, while central Norwegian institutions that by law have to take part in the hearing only submitted a simple approval of the suggested change to the legislation. An exception was the Norwegian Ministry of Defence, which delivered a more elaborated approval:

> Belonging to (*tilhørighet*) and having a good knowledge of Norwegian culture, language, tradition and Norwegian social relations (*samfunnsforhold*) is important to be able to look after community interests (*fellesskapsinteressene*) in Svalbard in a good way. (Norwegian Ministry of Defence 2021: 1, translation mine)

In other words, local knowledge of the history of Svalbard or the many specificities of life on the archipelago is redundant; transience, and the loss of knowledge and continuity mourned by the inhabitants (see Chapter 7) is not seen as a problem. At the same time, access to the kinds of knowledge the Norwegian Ministry of Justice and Public Security requires cannot be acquired in Longyearbyen. The most striking example of this is the absence of a Norwegian language course. In this way, Longyearbyen is not Norwegian enough. Norwegianness is 'an unequal resource' (Five Aarset 2018: 293).

Such a framing clashes on several scales with the reflection of Mari, one of the old-timers:

When LL was introduced, most of the people who lived here were against it. But there has never been any voting, we just got it introduced right away. Before you had the Svalbard Advisory Board, the government ruled over Svalbard, and the Board just came with advice. The difference from today is that with the Board there was stability. [...] I remember I talked once to a lawyer [...]. That was just when they put LL in place. And I told him: 'I am 100 per cent against LL.' He asked why and I said, 'How the hell are you going to have local democracy with all the moving in and out? When all the time you have new people coming in LL that move away after two years? There will be no stability and you won't get the right decisions made for the future of Longyearbyen. It will just cause frustration.' And he said, 'I haven't thought about this.' Then he went away, and a freelance journalist standing nearby started to laugh, came to me and asked whether I knew who he was. 'I don't,' I said. 'That was Carl August Fleischer,' she said: that is, one of the lawyers who designed Longyearbyen Lokalstyre, a very well-known advocate from Oslo. So exactly what we had feared then is coming true now. It think it's depressing to see what kind of decisions are being made. We should think how we want Longyearbyen to be in 20, 30 years. But most of those who live here think just 'now'. They don't bother about what comes in five years because they will move next year. Like that you don't make decisions from the bottom of your heart. But now we have Lokalstyre so we must make the best out of it. [...] We joke about it, us who live in Vestpynten and Bjørndalen. We say that we can't be bothered to live in town as it's too much hassle. Out there we live just like we used to live earlier. In a pleasant neighbourhood.

Mari's story about the lawyer from Oslo had a bitter taste. There was a hint of absurdity in the escape from the 'metropolis' bustling with life and swayed by shortsighted political decisions to the cape of Vestpynten or the valley of Bjørndalen, places too far from Longyearbyen to reach on foot but still within the reach of the electricity infrastructure. To live in a neighbourhood with *fellesskap*, one needs to retreat from all the construction work, overtourism, anonymity and transience to an area with a few cabins scattered around a badly maintained road. It reminded me of the trapper Harald Soleim, a solitary figure who overwintered 40 times on Svalbard on his own, and who fled from Tempelfjorden in the 1970s because there were too many tourists passing by on snowmobiles (Amundsen 2019). In every case, people like Mari who remember Longyearbyen governed exclusively from the outside, but based on advice given by a relatively stable advisory board where the few population segments were justly represented, feel both frustrated about not having been listened to and desperate about using the amputated local democracy as meaningfully as possible.

Towards the end of my fieldwork, voices were raised, mainly on Facebook and in the local newspaper, sarcastically (and some seriously) suggesting the abolition of local democracy and a return to full central governance. When most of the non-Norwegian residents were excluded from voting or being elected, the editor-in-chief (a new one after Hilde Kristin Røsvik, and on his way from Svalbard after about two years) published an editorial surfing on the town's wave of frustration:

> For 20 years, we have had a form of government where everyone with the right period of residence was allowed to vote. We can call it a democracy. Today, more than 2,500 people live here. Between 700 and 800 from 52 countries have another passport. When they now lose the opportunity to influence the society in which they live, we can no longer call this a democracy or a people's government. [...] Oslo decides everything anyway. It is a poorly concealed reality that this is done because Norway wants to strengthen the proportion of Norwegians in the archipelago. That is: to make it more difficult for everyone without a Norwegian passport to live here. [...] Be honest and acknowledge that democracy in Svalbard has been shot in the leg and hangs on a thin thread. The average length of residence is declining and it will be more difficult to recruit people for political positions. Those who settle here for a short period come to work and experience an adventure. Then the government can just as well take a step back, abolish local democracy, and take over the management of Longyearbyen. (Haugli 2022a)

In the commentary of Arne Holm, the journalist voiced his astonishment: 'Climate change is making it unbearably hot in the Arctic. At the human level, change triggered by politicians makes relations among people colder. Norwegian neo-nationalism on Svalbard arouses surprisingly little debate' (Holm 2022). Three years before the politics of exclusion became reified in the new rules for eligibility and voting rights, I met Ana Paula

Souza. Her story builds a bridge between a cooling social climate and the melting Arctic.

The Culture of Denial

> *Ana*: I came in the dark season. My first impression was it was a very warm, welcoming community.
>
> *Zdenka*: Have you changed your opinion on this?
>
> *Ana*: Very much. I understood I came in a moment of transition. My husband has been living here for 15 years and says the transition is very clear for him, too. [...] When you come to a place you arrive with expectations. And then you meet the reality of the place, and have to find balance there. [...] People come, interested in the city, excited, and then they are shocked it's something else.

Ana is a university-educated professional from Brazil, a psychologist and psychotherapist, used to volunteering and eager to contribute to what Olaf calls *fellesskap*. She was attentive not to present herself as somebody with a competence that was actually missing in town, as she was warned that newcomers who arrive and think they can change things from scratch hit a wall. After several unsuccessful tries when people never replied to her requests or when she felt ignored and marginalised, she started to wonder. Her impression was the idealistic *fellesskap* was not something anybody was welcome to foster; she was stopped by what she called 'the frontier', which she insisted was more than the language barrier. Neither LL, nor the hospital, nor the church were interested in her competence. The informed consent Ana signed had number 112. Two weeks later, I met my interlocutor number 119, Gunnar, a Norwegian old-timer who is open about his right-wing political views. His lifeworld was profoundly different from Ana's, but he felt an acute need to make me aware of the same issue that worried her, too: that people working for key local authorities create a sort of a closed group of people who can exercise power. Gunnar used the expression 'the howl of the tribe' (*stammens hyl*), referring to a mentality of 'a pack of wolves'. This shared perception of parallel, hierarchical and only instrumentally communicating structures within the town's population made people ruminate over how to 'crack the code' (*knekke koden*), as if there were specific rules or languages to master if one aspired to get access to those structures. When I asked Ana why she thought this was happening, she explained: 'I understood they are almost like ... mentally fighting for the territory. It's a way to keep hold of power.' In Ana's circle of friends, the term 'the Norwegians' did not apply to everybody of Norwegian nationality; the synonym, which Ana claimed other people used as well, was 'the mafia'. Ana had many friends among Norwegians who were concerned about growing barriers and impenetrable frontiers, and who would have liked to see a different, less excluding development of the town that used to be proud of being international while now this seems a shameful chapter in its recent history. 'The Norwegians' were thus an entity for Ana; agenda setters, gatekeepers, people in power and working in key institutions.

When I asked Ana about the term 'local community', she spoke about the same issues as many of my interlocutors already encountered throughout the book, or those whose voices are not heard in the text but who contributed to the tapestry woven with data created throughout the fieldwork. She was particularly scandalised by the ignorance of the international character of the town, manifesting itself, for example, in making important information only available in Norwegian, which – as I have shown in Chapters 6 and 8 – is something tour guides, Asian residents and others suffer from. 'We call it the culture of denial.' There is superdiversity, but the state strives, with the help of local authorities, to cover it up and deny its existence.

In her seminal work combining sociological and ethnographic methods, Norgaard (2011) explores in meticulous detail how the inhabitants of Bygdaby, an anonymous rural settlement in the Norwegian west coast, know about the troublesome issue of global warming, but struggle with knowing how to know about it, and how to tackle it. Denial seems to be a collective strategy used to respond to the overwhelming threat of climate change, but also a socially framed and reinforced pattern influencing narratives of culture, emotions and democracy. The culture of denial Ana describes in her story is linked to the practice of silencing discussion of questions of justice and human dignity, which makes large parts of the town's population feel invisible. Norgaard shows the connection between socially organised denial and democratic participation, and argues that large societal issues (such as climate change or inequality) are impossible to tackle in an atmosphere where the culture of denial dominates.

The social organisation of denial also raises questions about new limitations for democratic process in a world where the reorganisation of time and space mask the consequences of human actions. [...] The notion of socially organised denial places culture and emotion at the centre of social reproduction and legitimation. If we are concerned about climate change, the notion of socially organised denial clearly has implications for what to do next. (Norgaard 2011: 208)

When Norgaard shares her ideas about what to do next, she is careful not to offer universal one-size-fit-all solutions, and suggests that to make the path, one has to walk it. To be able to do so, the fundamental condition is local political renewal:

Participation [...] will reduce the gaps between abstract information and daily life, decrease the sense of a double reality, and bring home impacts in economic, infrastructural, and physical terms. [...] What is important for effectiveness, however, is that people understand and experience activities as more than ends in themselves. Local political renewal cannot be enough on its own. But it may be the important next step for individuals in breaking through the absurdity of the double life and for renewing democratic process. [...] As Eliasoph put it, 'with active, mindful political participation, we weave reality and a place for ourselves within it.' (Norgaard 2011: 208)

Similarly to communitification, the process of political and emotional mobilisation in order to co-design desired futures for the place one inhabits, overcoming the culture of denial, is a key factor on the way to more sustainable ways of life, in the Arctic and beyond. While Longyearbyen has grown and increased its ecological footprint, clinging to the logic of extractivism, it also got stuck in the practice of denying the lived experience of many who are part of its present story, but not its futuring. The *ouroboros* leads us not back to where we started, but it keeps returning to the uncomfortable question of how to live in a warming world where many desire what only few have. Put differently, how to decarbonise, turn away from extractivism and attend to injustices and inequalities in superdiverse societies. The culture of denial connects the fundamental paradox of managing the heavily politicised 'Svalbard wilderness' through extractivist industries financed by fossil economies, with the increasing social acceptance of practices and discourses legitimising exploitation and marginalisation of people who fail to serve the geopolitical purpose of Longyearbyen's existence.

Figure 4   Longyearbyen Green Future – Soon It's Here

Source: Courtesy of Polar Money News (PMN is a satirical
Facebook page run by anonymous Longyearbyen residents).

Conclusion:
The Paradox of Svalbard

From its early seventeenth-century origins in both international relations and international law, Svalbard has been a boxed off area on the map, a collection of islands contained within lines drawn across the sea. From the very start, Svalbard has been the object of attempted – and failed – possession by various states. [...] By the same token, Svalbard itself – Svalbard as a place – has often been less important in the international discourse than has Svalbard as a principle. It is this understanding of Svalbard that dominates the international discourse, including much of the scholarly discourse. Such an understanding at best grudgingly acknowledges the existence of those people who live in and visit Svalbard, focusing instead on the states that wish to somehow make Svalbard their own. [...] It is one thing to question why anyone really cares about a cluster of rocks in the Arctic, and it is another to question why the international discourse has come to conceptualize this particular cluster of rocks as a distinct territory, largely without reference to the thoughts, aspirations, and concerns of the people who actually live there. (Grydehøj 2020: 268)

My work in Longyearbyen had an ambition to attend to Grydehøj's call to acknowledge that there are people living on the cluster of pointy rocks in the Arctic ocean. *The Paradox of Svalbard* is an ethnographic study digging into multi-layered aspects of accelerated change that people in Longyearbyen manage to live with – despite the difficulties.

My project's mentor, the Norwegian social anthropologist Thomas Hylland Eriksen, is a strong voice reaching out beyond the academia arguing for understanding complexity as a way of seeing (Eriksen 2015). The narrative of Svalbard as a showroom for climate solutions is simplistic, while the stories I follow in the book are complex. In the introductory essay of their volume on climate, capitalism and communities, Stensrud and Eriksen (2019) argue that it makes sense to study 'how the effects of changes in the climate, weather and environment intersect with the effects of economic policies and social inequality' (2019: 1). This is also where I see the contribution of my ethnography – my hope is to offer a deeper insight into how climate change and globalisation are locally entangled, and what kind of inequalities and injustices are triggered by extractive governance over a place full of meanings and values.

> The double bind between growth and sustainability runs deep in contemporary modern societies; since growth and prosperity continue to be associated with high consumption of fossil fuels, it is difficult to reconcile the opposing ideals of economic growth and ecological sustainability (Stensrud and Eriksen 2019: 2–3).

Longyearbyen's local politicians picked their Sustainable Development Goals in 2021, to pursue in the years to come. The shopping list included the most contradictory one, SDG 8: Decent work and economic growth. While decent work is something many in Longyearbyen call for, the sticky idea of growth lurks around, casting doubts on where it is we are actually heading. As put by Harald, a non-resident scientist who regularly visits Svalbard whom I met at UNIS:

> The main problem is the way the economy is run. It is based on growth. But in the long run, everything cannot grow. I am a pessimist. Everybody talks about, you know, how to mitigate climate change, and do things about the climate, and CO_2 quotas and whatever, but when it comes to the show, there is little interest in doing anything about it. Even at the level of the Norwegian government. They still want to produce oil. And they don't want to put restrictions on people's car use or holidays by plane. And then they buy quota in some developing countries and say, ok, now we have bought these quota and we can continue drilling oil. If they want to reduce CO_2 emissions, what they do now doesn't work. It is very clear.

Seeing Svalbard as a miniature of Norway is paradoxical because it is true in a way rather different from the one those who propose this narrative would wish. In line with Anker's (2020) study of how Norway became a powerful periphery and environmental pioneer for the world, it is possible to look at Svalbard as a periphery of the periphery, an experimental territory for the nation's environmental endeavour. However, 'to follow the money of a rich nation on an environmental mission of bettering the world may not lead to the green results imagined' (Anker 2020: 238). Unveiling the aspects of fluid environments, extractive economies and disempowered communities, all hidden in the paradox of Svalbard, was my response to Stensrud and Eriksen's further call to 'address global questions about inequality, capitalism and environmental damage with a basis in ethnography' (2019: 5). This is done for the sake of 'reimagination and reworking of communities, societies and landscapes, especially those dominated by industrial capitalism' (2019: 9). Building on their claim that 'the rhetorical move from capitalism to green capitalism does not radically change the relationship between capital and nature, [with] sustainability as a powerful, historically produced discourse linked to the modern ideology of progress and development' (2019: 10), I traced how both Life and Nonlife (Povinelli 2016) are converted into commodities. To be able to do that, I had to make sense of scales of attention and analyse extractivism in its manifold forms. I felt inspired by the call:

> for more ethnographic studies of the impacts of climate change in terrains marked by social inequality, and also more nuanced analysis of how projects of environmental adaptation and mitigation are implemented in realities of inequality and difference. (Stensrud and Eriksen 2019: 15)

The Paradox of Svalbard is grounded in the lived experience of a place where Norwegians project their imagination about an ultimate space that is 'more Norwegian than Norway' in terms of an environment that invites exploration of the limits of human ventures, and at the same time is 'never Norwegian enough' in terms of a particular legal framework and recently also superdiversity that supposedly threatens its Norwegian identity. Or, in other words, it is 'part of Norway when it fits, and not when it doesn't' (Malmo 2021).

Grydehøj (2020: 280) engages with Svalbard's paradoxical nature in the following way:

> If Svalbard is not just a place of wild, untamed nature but a place of specifically *Norwegian* wild, untamed nature (Roberts and Paglia 2016), then activities undertaken in Svalbard by non-Norwegians can be either integrated into the narrative of a Norwegian Svalbard or challenged as inappropriate to the environment, depending on what is most convenient for the Norwegian state.

My ethnography shows the Norwegian state chooses the framing of inappropriateness.

The art of extraction seems to be limitless in Svalbard: we can extract life from the ocean, fossil fuels from the guts of the mountains or from the sea bed, but it is also possible to exploit memories, stories, knowledge, experience, and labour. But what is the ambition of ethnography? To describe how people are turned into resources, depending on the geopolitical context seen at one time as 'useful' and at another as 'meaningless', or are the ambitions higher than that? As Latour states ironically, ethnography is often content with 'a glorified version of story telling to which were added some radical pronouncements against power, injustice and domination' (Latour 2014: 139–AAA 8). I was hoping to do more than that. Can ethnography aspire beyond description and analysis, to mobilise action? What can I give back to Longyearbyen and its people? And how can I pay forward to them? As Singleton et al. (2021: 15) put it, I do not resign from 'reflecting upon … [my] role as an ally or at least a colleague in ongoing and future struggles. Research ends, but struggles continue.' In Svalbard, the call for non-extractive science (inclusive, participatory, transparent, co-productive, humble and reciprocal in the sense of giving back and paying forward) has been gaining attention recently, promoting a 'public science' that contributes more to the 'social life of the community' (Bravo 2006: 237). In spite of knowing that talking about the 'local community' is troublesome in Longyearbyen, being just another brick in Svalbard extractivism was not my plan.

Peeling Away the Layers of the Paradox of Svalbard: A Brief Guide to the Maze of the Book

In the introduction, I argued that the concept of the Anthropocene has aspects that increase the risk of it being appropriated as another fairy tale of teleological progress, which is apparent in the representation of Svalbard as a 'miniature of the world' becoming sustainable. The loops of the *ouroboros* touching upon different themes enable the reader to explore local meanings and manifestations of 'climate change' and 'globalisation'. While general fairy tales cover up the lived experiences of places, the *ouroboros* disentangles from them the big story and pays attention to meaningful spheres instead of a meaningless globe (Ingold 2021). The stories I tell, thanks to what my interlocutors shared with me, are neither linear nor cyclical. Through them, it is possible to unpack the many layers of the

paradox of Svalbard, not just 'the only story in town'. There are no minia-
tures of the world, but there are lived places and temporalities. Paradox is
an analytical tool I apply in my case study, but – as I hope comes through
the empirical material presented in each chapter – also an emic category.
People in Svalbard realise the manifold contradictions of the place. They
experience different forms, scales and rates of change, and the different
pressures and kinds of stress that emerge as a result of the intertwined
processes.

Chapter 1 shows how my path began, reflecting on time scales and
climate change with natural scientists, mostly geologists at UNIS. It
is possible to narrate very different fairy tales of change, with different
morals and ways of linking the deep past with alternative stories about
the future. Complications arise when different scales of attention clash.
I explore climate change in Longyearbyen in a relational way, communicated through stories stretching over diverse places and temporalities.
Ethnography can be seen as a 'fossilised ripple in the sand', ephemeral
snippets of life that perhaps no one expects to be documented, and yet
they are, inviting a conversation that digs deeply into the 'climate paradox'.

In Chapter 2, the reader meets the character of the avalanche as a game
changer, and the discourse of safety and responsibility as keywords in local
debates about climate change. I also discuss other ways of how changing
environments matter, for example, through altered mobility patterns
and a longing for the familiar. This nostalgia for an idealised past with
stable weather patterns, cold winters and dry summers has something in
common with the nostalgia for a less diverse neighbourhood; already in
the second chapter it becomes clear there are changes other than those in
the fluid environments that occupy people's minds. People's accounts of
observed changes bear witness to sense of place, and attachment to place,
being modified and negotiations about the future unfolding. The stories
are messy, enmeshed with people's values and identities, unlike the seemingly apolitical graphs and charts climate science presents to the public.

Chapter 3 winds up the first part of the book, shifting the focus from
what is observed, perceived and experienced to how the changes are
interpreted in the local discourse of climate change. I give examples of
counter-stories people tell to address the unspoken, the disturbing talk
about climate change in Longyearbyen where much more is going on
besides. In the issue of control over housing and the hostage of melting
glaciers, we are introduced to the backstage of the 'canary in the coal mine'

narrative. What and who is given precedence, what and who is being forgotten, and why?

Part II of the book is an excursion into the mechanisms of extractivism as it manifests in Longyearbyen undergoing a transition from coal
mining to the new industries of tourism, research and technological
development. In Chapter 4, I argue that tourism and science are kindred
extractivisms, considering 'what is out there' as possible resources for
further linear development. Also, more subtle phenomena such as experiences, memories and knowledge can be mined. The paradox of Svalbard
here entails the struggle to disentangle from the carbon utopia of the past
while embracing an extractivist post-carbon utopia of a future decided
elsewhere. In addition, while there are local aspirations to turn tourism
and science into non-extractive practices through looking for linkages,
confluences and giving-back modes of operation, interconnections are not
welcome and the state prefers to keep the spheres separate.

Chapter 5 is an exploration of how 'big powers' and 'little people' encounter each other in Longyearbyen, and how people negotiate the meanings
of their life stories when realising they have been serving agendas beyond
their reach or control. During the rapid economic diversification since
the 1990s, being small and local has not exactly proven to be an advantage
when competing with the big and global. Ethnography helps us understand what an economic transition of such a scale feels like from within.

The last chapter about extractive economies, Chapter 6, brings in a new
perspective into the discussion about (un)sustainable tourism. While the
environmental paradox of Svalbard with regard to the tourist industry is
obvious, social justice has been completely left out of the picture. The
paradox of Svalbard is lived on multiple levels by the tour guides, an international and politically marginalised group of disposable people who do
not easily fit the narrative of a Norwegian hub of climate solutions. In
their struggles, the reader can hear the nationalistic undertone often
evident in the fairy tale about a showroom of sustainability.

In Part III, further paradoxes await reflection. Chapter 7 dwells on
the trouble of the 'local community', which from within is perceived as
'un-Norwegian' even while the politics of difference practised in town is
what makes it typically Norwegian. People mourn an idealised past but
move forward to an active future-making they cannot affect as the process
of political mobilisation is hindered.

In Chapter 8, I visit the lifeworlds of people who have been part of
the town's (hi)story for almost 30 years, but they match the state's nar-

rative about 'the neighbourhood' just as poorly as the international tour guides. The decreasing social cohesion that my interlocutors comment on in Chapter 7 manifests in the shrinking margins of inclusion for those who are 'too many and too different' – such as the Thais and Filipinx of Longyearbyen. The politics of exclusion, pursued by the state with an ever fiercer effort, causes unhealthy dependencies and growing inequalities.

This part of the book on disempowered communities ends with Chapter 9, which is on the 'culture of denial'. Denying Longyearbyen's superdiversity reconnects my material with the problem of climate change, which can only be solved through a process of political renewal. Such a process necessitates a divorce from the hypocrisy and absurdity of living a double life in denial, nourished by the subtle bureaucratic violence of a supposedly egalitarian – and highly fossil-dependent – state.

Making Arctic Places Green, or Just, or Both? Real Life in a 'Mythical Country'

> The Arctic has never been an inert, empty space, despite what many popular stereotypes of it as timeless, vast frozen wilderness otherwise suggest. A region of flux, impermanence and becomings, it has experienced tremendous shifts in climate over millennia. [...] Many parts of the Arctic are attracting increased interest as sites for extractive industries, cruise ship tourism is expanding into areas previously thought of as remote and difficult to navigate through, especially in places where ice has usually impeded maritime transit, and conservation organisations and environmentalist groups place iconic species such as polar bears and whales up front and centre in global campaigns to save and protect circumpolar lands, waters and ice. [...] The challenges arising from environmental transformation combine with political and economic ones, as well as broader processes of social change. (Nuttall 2019: 57–8)

The Paradox of Svalbard, besides being about climate change and globalisation, is also a book about an Arctic place. A focus on making Arctic places (more) sustainable is increasing while the scramble for the poles is intensifying – there are tempting opportunities for shipping and resource extraction in an ice-free Arctic. In May 2022, *Svalbardposten* (Haugli 2022b) informed its readers that LNS (Leonhard Nilsen & Sønner, a construction and logistics company with a Svalbard branch) might start writing a new chapter in the story of Arctic extractivisim. The world's northernmost zinc and lead mine in Citronen Fjord in North Greenland, at 83° North, will open in 2023 in collaboration with the Australian Ironbark Zink Ltd. According to Wikipedia, the People's Republic of China is also involved in the mining plans. The miners are supposed to work in Citronen Fjord and live in Longyearbyen, only a few hours away by plane (about 600 miles). Such are the times of the Arctic double bind: the Arctic amplification warms it up faster, but also the extractive economies keep spinning, contributing to a further acceleration of environmental change that in many places, including those in the Arctic, means a crisis. According to Nixon (2011), change is a cultural constant (see also Kaltenborn et al. 2020), but its pace is not. In his concept of slow violence, Nixon (2011) looks at forms of displacement that do not require physical movement away from places of belonging; he speaks about communities 'stranded in a place stripped of the very characteristics that made it inhabitable' (2011: 18). While what is happening in Svalbard does not at first sight resemble the loss of Arctic communities who are being deprived of their lands and resources beneath their feet by extractive industries (hard or soft) embedded in capitalism, a sense of loss is palpable there, too. It is a loss that might seem rather unspectacular to some; how can such a transient and non-Indigenous settler community as that of Longyearbyen relate to the pains of the people in Kalaallit Nunaat or Nunavut? What makes Longyearbyen relevant is what we can learn from its particular Arctic story about the '200-year experiment in hydrocarbon-fueled capitalism whose historic beneficiaries have been disproportionately rich and white' (Nixon 2011: 266). This experiment continues in Longyearbyen, the living paradox of the twenty-first century: how to live in a warming world where many desire what only few have? Or should we rather say, where all deserve what only a few keep for themselves? This is how one of Ana's psychotherapy clients described the showroom, exposing its vulnerabilities and politics of exclusion during the pandemic: 'As if we're inside the *Titanic* and it is about to sink down in a cold dark place without a safety vest for everyone.' How inspiring are unjust places that build their green narrative on representing nature as vulnerable while forgetting about people? As Hastrup (2019: 42) writes:

> the view of the Arctic landscape has changed from being powerful and even hostile, to being fragile and in need of protection. Without understanding the long-term processes of colonisation, resource exploitation

and military logic, we may not be able to see the immensity of what is happening now.

While the Svalbard Treaty soothes its readers by stating that Svalbard is an area that cannot be used for any activity serving military purposes, the way Russia (for example, as of 2022) threatens stability in various regions of the world, including the Arctic, puts people and states on the alert. Is peace in and around Svalbard at risk, with the Russian settlement of Barentsburg and Russian officials regularly claiming that Norway's interpretation of the Svalbard Treaty is misleading, and that their equal rights to come and do business disrespected? This is not a question I could answer with the help of ethnography, but I argue that international relations uninformed by meticulous understanding of the local context have alienating impacts. There is always an advantage in understanding what is actually happening in the field, on the condition we believe that all human lives are meaningful.

During the Covid-19 pandemic, when travelling was regulated by 'traffic lights' according to the level of the infection spreading rate, I had to fly to Oslo. When I arrived to the passport control at Oslo Airport, I was – obviously – coming from outside the Schengen Area.

Officer: So you live in...?

Zdenka: I live in Svalbard.

Officer: Oh! [to her colleague] She says she lives in Svalbard. She speaks Norwegian. But she has a Czech passport. Is Svalbard a green area?

The other officer: Well ...

Officer: But isn't Svalbard part of Norway?

The other officer: Well ...

Officer: [to me] What are you up to in Norway?

Zdenka: Ehm ... I am on a business trip until Sunday, then I go back.

Officer: Do you have a ...?

Zdenka: [eager to satisfy her] A confirmation of residence?

Officer: No, a confirmation that you are coming for a business trip.

A few embarrassing minutes followed. I thought I should be allowed to do whatever, but I managed to dig out an email confirming there is a purpose to my trip. While I was sweating, the officer continued:

Officer: Svalbard. A mythical country really. Why don't they issue ID cards for the residents, I wonder ...

I found the conversation amusing (when I was finally allowed through passport control). Svalbard is not a country, and it has been part of the officer's for more than hundred years. Its residents have no ID cards because any measure strengthening attachment to the place might also contribute to a stronger local identity, thus ownership and willingness to actively shape the place's future. These layers still seem to be hidden behind a mythical veil for most of the Norwegian public, unaware of the struggles and dilemmas people experience in this particular place. An unreal place where people live real lives under the aurora borealis. In this respect, the otherwise rather atypical Longyearbyen fits Nuttall's (2019) comment on how overheating translates into Arctic realities: 'For those who live there, and who seek to have their voices heard [...] the region has become characterised by uncertainty, as well as subject to a hardened global gaze, in a way they have not really experienced before' (2019: 69).

Grydehøj (2020: 280) argues for careful attention scaling:

It is thus not a question of whether Svalbard should be understood through the locally grounded perspective of the people who live and visit there or through the more abstract perspective of international relations. Neither Svalbard's community life nor its importance within Arctic international relations can be understood independently, without reference to the other.

In *The Paradox of Svalbard*, I document how local (hi)stories matter, and how without listening for them, we risk taking decisions that fail to do justice to the places and people that will be impacted by them. 'Nothing about Svalbard is sustainable,' a tour guide once expressed what I have been reflecting on many times. In the interview with Henrik, the miner, he confessed:

If you were really strict about [these environmental] ideas then we shouldn't be here at all. [laughter] But we live here because there must be people living here, because of this interest sphere. The Norwegians are worried to lose to the Russians and Chinese and Americans and so on. [...] It's easy to squeeze them because the others are so big. We have such a low tax here to make it interesting for Norwegians to come here. [...] [T]he Norwegian government [is] trying to hide it but they're really worried about getting too many foreigners here. You can say we are a mini version of the Norwegian society.

There is a rhetorical question circulating in Longyearbyen. It is a resigned remark used in situations when the environmental, economic and social paradox of Svalbard manifests itself: 'Shouldn't we all just leave?' The problem is, there is no 'just leaving', neither Svalbard, nor the Earth. There is, however, a lot of unjust leaving out and leaving behind. If we can tackle those, we might be able to abandon the anti-humanistic idea of leaving and focus on living.

... and the Ouroboros Winds Up: Fairy Tales and Futuring

The difference between a fairy tale and a story is that fairy tales hardly ever have an open ending; they finish with a straightforward message that determines how the tale is narrated. Stories can be chaotic, confusing and contradictory, and sometimes the only 'message' we can distil from them is the question they make us ask but to which they give no definite answer.

The fairy tale about Svalbard as a canary in the coal mine that is now going green is trying to convince us that we can tackle ecological ruination without abandoning economic growth, and without attending to issues of disempowerment, exploitation and alienation. It flirts with 'fantasies about improved collective conditions while being increasingly indifferent to the structural violence supporting this economy' (Masco 2015: S65). The showcase of both the crisis and its solutions promises we can change while staying the same (Povinelli 2016).

There are indeed too many contradictory opinions, processes, scales, actors, and conflicts at work for this to take place as a unified or unifying transitory narrative. It is perhaps this seemingly contradictory co-existence that defines post-carbon utopia and the drama of Anthropogenic socio-natures in Svalbard – that is, human attempts to reconnect with, or recreate, natural wilderness while simultaneously seeking imaginary solutions to continue life as we (or, some of us) know it. (Ødegaard 2022: 15–16)

There has been quite a bit of debunking (Latour 2004) in documenting and unpacking the paradox of Svalbard. But what about caring? How can we pursue alternative pathways? How can communities nudge themselves in 'a different, and potentially more sustainable, direction' (Eriksen and Selboe 2015: 131)? The green transformation is another boundary object (Amundsen and Hermansen 2021) and the only way to make it more tangible while facing chronic uncertainty is to emplace it.

> Including situated knowledge and the lived experiences of place in response to place change are thus critical steps towards democratising knowledge creation and adjusting actions for configuring places. (Manzo et al. 2021: 341)

Instead of returning to nature, I would like to propose returning to politics. Politics that is reflective, able to 'recognise and dismantle place injustices, continuous (dis)empowerment, (un)sustainabilities and local (mal)adaptations' (Manzo et al. 2021: 343). Who can 'crack the code' and disentangle Svalbard from its many paradoxes? Not the state, neither capitalism, nor modernism, but maybe political empowerment, imagination and hope. Hope is:

> not a belief that everything will be fine; it is about developing broad perspectives with specific possibilities: hope locates itself in the premises that we don't know what will happen and that in the spaciousness of uncertainty is room to act. (Manzo et al. 2021: 344)

I resonate with Head's (2015) reasoning that the time of decarbonisation is coming sometime between 'now' and 'very soon', and it will be a dramatic move, either by force or by choice. It might not necessarily be a catastrophe, though; we would, however, need to bid the modernist future farewell. So far, Svalbard seems stuck. Many of the local voices I heard throughout my stay in Longyearbyen, however, expressed hope for a non-utopian yet positive future and willingness to work for it, had they only been given the chance – or had there been any potential for creating the chance themselves. What is needed is not a streamlined utopia designed by the state, but rather the use of 'practical knowledge, informal processes, and improvisation in the face of unpredictability' (Scott 1998: 6). The search and the struggle is much more interesting than the showroom.

Afterword

Hilde Henningsen

There are many voices in Longyearbyen, yet they are not all heard ...
There are many stories in Longyearbyen, yet they are not all told ...

This book reveals a variety of complex stories emerging from encounters with a cross-section of individuals in the town that has been my home for over 35 years, thus shedding light on untold aspects and facets of the diamond – or the multi-headed troll – that makes this community what it is.

There is a multitude of scientific reports on nature in our Arctic region, but not so many in the fields of sociology and anthropology. Simultaneously there is a tiredness, almost fatigue, among some of the old-timers to be asked to yet again to answer the same endless questions about life in our city. Over the past decades, we have entered the era of reality shows; *Livet er Svalbard*, *Life on the Edge*, *Kompani Spitsbergen*, social media reveals and marketing, Instagram stories, travel blogs and the vast TikTok universe depicting our community as it spins towards the future.

We may think ours is a local community, a local arena, and it is – but it is also so much more a national and international, or rather geopolitical scene, whether we like it or not. We are currently living amidst the most significant turn of the tide – a massive paradigm shift – in the history of our town. Our community is ever changing, with a sizeable turnover that some say results in a lot more consumers than there are contributors.

I heard of Zdenka long before I met her; *there is this woman, she is probably just another journalist, filmmaker, author, TV reporter, whatever ... a social anthropologist who wants to meet with people ... and social science is not really our thing ...* And I realised, finally, it is happening! Someone seriously wants to study people, relations, discourse, interactions with nature, culture, climate, the whole shebang ... and not just every minute scientific aspect of a rock, an organism, a body of water or mud.

As it turned out, many spoke to her, rather *despite* than *because of*. Zdenka challenged, she questioned, she listened, and she herself was challenged by the responses she got, yet she was brave enough to reflect upon them and the consequences they encompassed.

I have a strong feeling that she has depicted the layers of the Longyearbyen community with far more insight than many of our visiting politicians, experts, and consultants.

It is not so hard to come somewhere with an assumption and find 'proof' that it is somehow right or mostly so. But Zdenka didn't just listen to what she wanted to hear, but rather she listened to all that could be heard. Zdenka did not just coldly observe, but warmly participated. She did not just stand on the outside looking at, but lived the life within, among us locals, husband, children, her family – all that. She lived the life of a Longyearbyen citizen (defined by the local register as 'having the intention to stay six months or more'). Just like the rest of us, she brought her life to Longyearbyen, and lived it here; she ate, she slept, she struggled, she froze, she rejoiced, she protested, she talked, she listened. Possibly her warm participation was her biggest challenge in the role of the scientist, because her heart was in there, as much as her head.

She also utilised her secret weapon, her incredible ability to adapt and learn the language of the land, the key to communication, literacy, trust, credibility and sincerity.

Language is the tool by which people connect and unite, yet it is also the provider of words that divide and conquer. Zdenka is a female Askeladden,[1] who observed and utilised the perceived knowledge and taught the rest of us a lesson. She is also the scientist who braved the stormy waters of entanglement in a small community, striving to stay afloat on her anthropological raft.

Zdenka has given voice to unheard voices in the community by creating arenas of common ground – community dialogues. She brought people together across group boundaries.

Zdenka shattered 'the Svalbard bubble' – not trashing it, but splitting it into its various pieces, and then worked hard to put them together again within an explanatory context. She strived to look upon contradictions and paradoxes as aspects of the same reality, rather than truths and untruths. In a layered community with so many parallel subcultures, she combined information in a grid, seeking order in chaos.

In one of the community dialogues an old-timer said we need not forget Why We Are Here: We may have come here by our own means, will and

---

1.   Askeladden is a male character in several well-known Norwegian folk tales, a loser at first sight, but with time he succeeds in difficult tasks.

decision, but we stay here within the possibilities and limitations defined for us by our government. We are the tools by which they govern. We may not like it. We may not want it. Yet we are part of the 'current wrapping' to uphold the Norwegian community in Svalbard.

Zdenka has challenged me, pushed me to listen more, to reflect more, to recalculate and reconsider, to seek understanding, and look beyond what first meets the eye. She has uncovered and challenged my prejudices.

I believe this book has the power to play a significant part in future decision making and public discourse, and uncover a greater understanding that the basis for decision making should be far more multifaceted than it has been so far.

I strongly believe that the more balanced and thorough information going into the calculation, the better the solutions. Failure to acknowledge or trying to exclude parts of the phenomena does not help.

Zdenka writes that 'Local (hi)stories matter, and without listening for them, we risk taking decisions that look great on paper (or in the media) but fail to do justice to places and people impacted.'

She quotes that 'To make the path one has to walk it.' And 'with active, mindful political participation, we weave reality and a place for ourselves within it'.

'Nothing about Svalbard is sustainable.' So, should I stay or should I go? Or should we live the life that makes us Come Together, Right Now?

Who would have known how precise Zdenka's timing was for unravelling these paradoxes caught in this turmoil of rapid change!

References

Abram, S. (2018). *Likhet* is not equality: Discussing Norway in English and Norwegian. In: Bendixsen, S.K.N., Bringslid, M.B. and Vike, H. (eds) *Egalitarianism in Scandinavia: Historical and Contemporary Perspectives*, pp. 87–108. Cham: Palgrave Macmillan.

Adomaitis, N. (2022). Oil firms must step up exploration off Norway to unlock potential – NPD. *Reuters*, 25 August. At: www.reuters.com/markets/commodities/oil-firms-must-step-up-exploration-off-norway-unlock-potential-npd-2022-08-25 (accessed 25 November 2022).

AECO (Association of Arctic Expedition Cruise Operators) (2022). *Field Staff Training, Education and Experience: Survey Results – Field Staff and Operators' Perspectives*. Tromsø, Norway: AECO.

Albrecht, G. (2005). 'Solastalgia': A new concept in health and identity. *PAN* 3: 41–55.

Amit, V. and Rapport, N. (2002). *The Trouble with Community: Anthropological Reflections on Movement, Identity and Collectivity*. London: Pluto Press.

Amundsen, B. (2019). *Harald: Førti år elene på Svalbard*. Bergen: Vigmostad & Bjørke AS.

Amundsen, H. (2015). Place attachment as a driver of adaptation in coastal communities in Northern Norway. *Local Environment* 20(3): 257–276.

Amundsen, H. and Hermansen, E.A. (2021). Green transformation is a boundary object: An analysis of conceptualisation of transformation in Norwegian primary industries. *Environment and Planning E: Nature and Space* 4(3): 864–885.

Andersen, T. (2022). Negotiating trade-offs between the environment, sustainability and mass tourism amongst guides on Svalbard. *Polar Record* 58: E9.

Anderson, B. (2016 [1983]). *Imagined Communities: Reflections on the Origin and Spread of Nationalism*. London: Verso.

Anker, P. (2020). *The Power of the Periphery: How Norway Became an Environmental Pioneer for the World*. Cambridge: Cambridge University Press.

Anonymous (2022). Norwegian Arctic archipelago feels the heat of climate change. *Thelocal.no*, 22 June. At: www.thelocal.no/20220622/norwegian-arctic-archipelago-feeling-the-heat-of-climate-change/ (accessed 5 June 2021).

Arlov, T.B. (2003). *Svalbards historie*. Trondheim: Tapir akademisk forlag.

Arlov, T.B. (2020a). Maps and geographical names as tokens of national interests: The Spitsbergen vs. Svalbard case. *Nordlit* 45: 4–17.

Arlov, T.B. (2020b). The interest game about Svalbard 1915–1925. Presentation at a seminar on 'The Svalbard Treaty: 100 Years', Governor of Svalbard in collaboration with Svalbard Museum and UNIS, 4 February.

Avango, D., Lépy, E., Brännströmm, M., Heikkinen, H.I., Komu, T., Pashkevich, A. and Österlin, C. (2023). Heritage for the future: Narrating abandoned mining sites. In: Sörlin, S. (ed.) *Resource Extraction and Arctic Communities: The New Extractivist Paradigm*, pp. 206–228. Cambridge: Cambridge University Press.

Bårdseth, A. (2020). Tredjelandsborgere: 20 fikk reisetilskudd. *Svalbardposten*, 17 November. At: https://svalbardposten.no/nyheter/20-fikk-reisetilskudd/19.13221 (accessed 5 June 2021).

Bårdseth, A. (2021a). Utenlandske familier valgte å bli på Svalbard av hensyn til barna. *Svalbardposten*, 1 March. At: https://svalbardposten.no/nyheter/utenlandske-familier-valgte-a-bli-pa-svalbard-av-hensyn-til-barna/19.13651 (accessed 5 June 2021).

Bårdseth, A. (2021b). Krisesenter slår alarm: – Svalbard fristed for voldsmenn. *Svalbardposten*, 19 April. At: https://svalbardposten.no/nyheter/svalbard-fristed-for-voldsmenn/19.13861 (accessed 15 October 2021).

Barnes, J.A. (1954). Class and committees in a Norwegian island parish. *Human Relations* 7(1): 39–58.

Barnes, J., Dove, M., Weiss, H., Yager, K., Lahsen, M., Mathews et al. (2013). Contribution of anthropology to the study of climate change. *Nature Climate Change* 3(6): 541–544.

Bauman, Z. (2001). *Community: Seeking Safety in an Insecure World*. Cambridge: Polity Press.

Bissat, J.G. (2013). Effects of policy changes on Thai migration to Iceland. *International Migration* 51(2): 46–59.

Blackstock, K., White, V., McCrum, G., Scott, A. and Hunter, C. (2008). Measuring responsibility: An appraisal of a Scottish National Park's sustainable tourism indicators. *Journal of Sustainable Tourism* 16(3): 276–297.

Bravo, M.T. (2006). Science for the people: Northern field stations and governmentality. *British Journal of Canadian Studies* 19(2): 221–245.

Brode-Roger, D..(2021). Starving polar bears and melting ice: How the Arctic imaginary continues to colonize our perception of climate change in the circumpolar region. *International Review of Qualitative Research* 14(3): 497–509.

Brode-Roger D. (2023). The Svalbard Treaty and identity of place: Impacts and implications for Longyearbyen, Svalbard. *Polar Record* 59: E6.

Brode-Roger, D., Zhang, J., Meyer, A. and Sokolíčková, Z. (2022). Caught in between and in transit: Forced and encouraged (im)mobilities during the Covid-19 pandemic in Longyearbyen, Svalbard. *Geografiska Annaler, Series B: Human Geography*.

Bruun, M.H., Jakobsen, G.S. and Krøijer, S. (2011). Introduction: The concern for sociality. *Social Analysis* 55(2): 1–19.

Buckley, R. (2012). Sustainable tourism: Research and reality. *Annals of Tourism Research* 39(2): 528–546.

Büscher, B. and Davidov, V. (2016). Environmentally induced displacements in the ecotourism–extraction nexus. *Area* 48(2): 161–167.

Butler, R. (1999). Sustainable tourism: A state-of-the-art review. *Tourism Geographies* 1(1): 7–25.

Butler, R. (2018). Sustainable tourism in sensitive environments: A wolf in sheep's clothing? *Sustainability* 10(6): 1789.

Byström, J. (2019). *Tourism Development in Resource Peripheries: Conflicting and Unifying Spaces in Northern Sweden*. Unpublished Doctoral thesis. Umeå: Umeå University.

Carey, M. (2008). The politics of place: Inhabiting and defending glacier hazard zones in Peru's Cordillera Blanca. In: Orlove, B., Wiegandt, E. and Luckman, B. (eds) *Darkening Peaks: Glacier Retreat, Science and Society*, pp. 229–240. Berkeley: University of California Press.

Carrington, D. (2017). Arctic stronghold of world's seeds flooded after permafrost melts. *The Guardian*, 19 May. At: www.theguardian.com/environment/2017/may/19/arctic-stronghold-of-worlds-seeds-flooded-after-permafrost-melts (accessed 1 July 2022).

Carson, R. (2015 [1962]). *Silent Spring*. London: Penguin.

Chakrabarty, D. (2017). The politics of climate change is more than the politics of capitalism. *Theory, Culture & Society* 34(2–3): 25–37.

Chekin, L.S. (2020). Svalbarðs fundr. The place name Svalbard and its connotations in medieval and modern literature and cartography. *Nordlit* 45: 18–38.

Christian, M. (2017). Protecting tourism labor? Sustainable labels and private governance. *GeoJournal* 82(4): 805–821.

Cohen, A. (1985). *The Symbolic Construction of Community*. London: Ellis Horwood and Tavistock Publications.

Crate, S. (2008). Gone the bull of winter? Grappling with the cultural implications of and anthropology's role(s) in global climate change. *Current Anthropology* 49(4): 569–595.

Creed, G. (2006). *The Seductions of Community: Emancipations, Oppressions, Quandaries*. Santa Fe, TX: School of American Research.

Cruikshank, J. (2005). *Do Glaciers Listen? Local Knowledge, Colonial Encounters, and Social Imagination*. Vancouver, BC: UBC Press.

Dallmann, W., Blomeier, D., and Elvevold, S. (2015). *Geoscience Atlas of Svalbard*. Tromsø: Norsk polarinstitutt.

Dodds, K. and Nuttall, M. (2016). *The Scramble for the Poles: The Geopolitics of the Arctic and Antarctic*. Cambridge: Polity Press.

Dvorak, Z. (2019). Bærekraftig turisme I Arktis? En studie av utviklingen på Svalbard. Unpublished Master's thesis. Bodø: Nord universitet.

EEA (European Environment Agency) (2017). Projected population trends in the Arctic, 14 June. At: www.eea.europa.eu/data-and-maps/daviz/projected-population-trends-in-the-arctic#tab-chart_1 (accessed 5 June 2021).

Eriksen, S. and Selboe, T. (2015). Transforming toward or away from sustainability? How conflicting interests and aspirations influence local adaptation. In: O'Brien, K.L. and Selboe, E. (eds) *The Adaptive Challenge of Climate Change*, pp. 118–139. New York: Cambridge University Press.

Eriksen, T.H. (1993). *Typisk norsk: Essays om kulturen i Norge*. Oslo: C. Huitfeldt Forlag.

Eriksen, T.H. (2015). Cultural complexity. In: Vertovec, S. (eds) *Routledge International Handbook of Diversity Studies*, pp. 371–378. London: Routledge.

Eriksen, T.H. (2016). *Overheating: An Anthropology of Accelerated Change*. London: Pluto Press.

Eriksen, T.H. (2018). *Boomtown: Runaway Globalisation on the Queensland Coast.* London: Pluto Press.

Eriksen, T.H. (2022). Everything is at a time and in a place. Guest editorial. *Anthropology Today* 38(3): 1–2.

Eriksen, T.H. and Schober, E. (2016). *Identity Destabilised: Living in an Overheated World.* London: Pluto Press.

Feld, S. and Basso, K.H. (1996). *Senses of Place.* Santa Fe, TX: School of American Research Press.

Five Aarset, M. (2018). Conditional belonging: Middle-class ethnic minorities in Norway. In: Bendixsen, S.K.N., Bringslid, M.B. and Vike, H. (eds) *Egalitarianism in Scandinavia: Historical and Contemporary Perspectives,* pp. 291–311. Cham: Palgrave Macmillan.

Forbrukerradet.no (2017). Forbrukerrådet på plass på Svalbard, 13 December. At: www.forbrukerradet.no/siste-nytt/forbrukerradet-pa-plass-pa-svalbard/ (accessed 1 July 2022).

Fraser, A. (2019). On the front lines of climate change in the world's northernmost town. *Reuters,* 3 September. At: www.insider.com/on-the-front-lines-of-climate-change-in-the-worlds-northernmost-town-2019-9 (accessed 1 July 2022).

Fukuyama, F. (1992). *The End of History and the Last Man.* New York: Free Press.

Geyman, E.C., van Pelt, W.J.J., Maloof, A.C., Aas, H.F. and Kohler, J. (2022). Historical glacier change on Svalbard predicts doubling of mass loss by 2100. *Nature* 601: 374–379.

Goldberg, M.L. (2011). A chilling warning from a 'hot spot' in climate change. *UN Dispatch,* 29 June. At: www.undispatch.com/a-chilling-warning-from-a-%E2%80%9Chot-spot%E2%80%9D-in-climate-change/ (accessed 1 July 2022).

Graham, N. (2020). Fossil knowledge networks: Science, ecology, and the 'greening' of carbon extractive development. *Studies in Political Economy* 101(2): 93–113.

Granås, B. (2012). Ambiguous place meanings: Living with the industrially marked town in Kiruna, Sweden. *Geografiska Annaler. Series B, Human Geography* 94(2): 125–139.

Grimwood, B. (2015). Advancing tourism's moral morphology: Relational metaphors for just and sustainable Arctic tourism. *Tourist Studies* 15(1): 3–26.

Grydehøj, A. (2020). Svalbard: International relations in an exceptionally international territory. In: Coates, K.S. and Holroyd, C. (eds) *The Palgrave Handbook of Arctic Policy and Politics,* pp. 267–282. Cham: Palgrave Macmillan.

Gudeman, S.F. and Rivera, A. (1990). *Conversations in Colombia: The Domestic Economy in Life and Text.* Cambridge: Cambridge University Press.

Guía, J. (2021). Conceptualizing justice tourism and the promise of posthumanism. *Journal of Sustainable Tourism* 29(2–3): 503–520.

Gullestad, M. (1984). *Kitchen-table Society: A Case Study of the Family Life and Friendships of Young Working-class Mothers in Urban Norway.* Oslo: Scandinavian University Press.

Gullestad, M. (2002). Invisible fences: Egalitarianism, nationalism and racism. *Journal of the Royal Anthropological Institute* 8(1): 45–63.

Hacquebord, L. and Avango, D. (2009). Settlements in an Arctic resource frontier region. *Arctic Anthropology* 46(1–2): 25–39.

Hacquebord, L., Steenhuisen, F. and Waterbolk, H. (2003). English and Dutch whaling trade and whaling stations in Spitsbergen (Svalbard) before 1660. *International Journal of Maritime History* 15(2): 117–134.

Hadot, P. (2006). *The Veil of Isis: An Essay on the History of the Idea of Nature.* Cambridge, MA: Harvard University Press.

Hanrahan, M. (2017). Enduring polar explorers' Arctic imaginaries and the promotion of neoliberalism and colonialism in modern Greenland. *Polar Geography* 40(2): 102–120.

Hanssen-Bauer, I., van Pelt, W.J.J., Maloof, A.C., Aas, H.F. and Sorteberg, A. (2019). *Climate in Svalbard 2100 – A Knowledge Base for Climate Adaptation.* Oslo: Norwegian Environmental Agency.

Haraway, D. (2016). *Staying with the Trouble: Making Kin in the Chthulucene.* Durham, NC: Duke University Press.

Haraway, D., Ishikawa, N., Gilbert, S.F., Olwig, K., Tsing, A.L. and Bubandt, N. (2016). Anthropologists are talking – about the Anthropocene. *Ethnos* 81(3): 535–564.

Hastrup, K. (2013a). Scales of attention in fieldwork: Global connections and local concerns in the Arctic. *Ethnography* 14(2): 145–164.

Hastrup, K. (2013b). The ice as argument: Topographical mementos in the High Arctic. *Cambridge Anthropology* 31(1): 51–67.

Hastrup, K. (2019). A community on the brink of extinction? Ecological crises and ruined landscapes in northwest Greenland. In: Stensrud, A. and Eriksen, T. (eds) *Climate, Capitalism and Communities: An Anthropology of Environmental Overheating,* pp. 41–56. London: Pluto Press.

Hastrup, K. and Olwig, K. (2012). *Climate Change and Human Mobility: Global Challenges to the Social Sciences.* Cambridge: Cambridge University Press.

Haugli, B. (2022a). Farvel, demokrati. *Svalbardposten,* 24 June. At: www.svalbardposten.no/demokrati-politikk-stemmerett/farvel-demokrati/486812 (accessed 1 July 2022).

Haugli, B. (2022b). LNS skal drifte verdens nordligste gruve. *Svalbardposten,* 25 May. At: www.svalbardposten.no/gruvedrift-gronland-lns/lns-skal-drifte-verdens-nordligste-gruve/484979 (accessed 1 July 2022).

Head, L. (2015). The Anthropoceneans. *Geographical Research* 53(3): 313–320.

Helgesen, L., Holmén, K., Misund, O., Jenssen, E. and Holmén, J. (2015). *The Ice is Melting: Ethics in the Arctic.* Bergen: Fagbokforlaget.

Helgesen, L., Holmén, K. and Sjøbu, A. (2020). *Vårt frosne vann: Etiske refleksjoner når isen smelter.* Bergen: Fagbokforlaget.

Herva, V.-P., Varnajot, A. and Pashkevich, A. (2020). Bad Santa: Cultural heritage, mystification of the Arctic, and tourism as an extractive industry. *Polar Journal* 10(2): 375–396.

Higgins-Desbiolles, F., Doering, A. and Chew Bigby, B. (2022). Introduction – Socialising tourism: Reimagining tourism's purpose. In: Higgins-Desbiolles, F., Doering, A. and Chew Bigby, B. (eds) *Socialising Tourism: Rethinking Tourism for Social and Ecological Justice,* pp. 1–21. London: Routledge.

Hokholt Bjelland, K. (2019). *Naturguidens rolle i et bærekraftig reiseliv på Svalbard.* Unpublished Bachelor's thesis. Tromsø: UiT – Norges arktiske universitet.

Holm, A. (2022). Svalbard: While temperatures rise, relations between people grow cold. *High North News*, 24 June. At: www.highnorthnews.com/en/svalbard-while-temperatures-rise-relations-between-people-grow-cold?fbclid=IwAR3NFBNm1oYsKfx_obmItjmYcP544cVx_Kpl7oKtDuZl2qFB_Jw2GvgU1jQ (accessed 1 July 2022).

Hønneland, G. (1998). Compliance in the fishery protection zone around Svalbard. *Ocean Development and International Law* 29(4): 339–360.

Hovelsrud, G.K., Kaltenborn, B.P. and Olsen, J. (2020). Svalbard in transition: Adaptation to cross-scale changes in Longyearbyen. *Polar Journal* 10(2): 420–442.

Hovelsrud, G., Veland, S., Kaltenborn, B., Olsen, J. and Dannevig, H. (2021). Sustainable tourism in Svalbard: Balancing economic growth, sustainability, and environmental governance. *Polar Record* 57: E47.

Hulme, M. (2009). *Why We Disagree about Climate Change: Understanding Controversy, Inaction and Opportunity*. Cambridge: Cambridge University Press.

Ingold, T. (2007). *Lines: A Brief History*. London: Routledge.

Ingold, T. (2021). *The Perception of the Environment: Essays on Livelihood, Dwelling and Skill*. London: Routledge.

Ingold, T. and Simonetti, C. (2022). Introducing solid fluids. *Theory, Culture & Society* 39(2): 3–29.

Ingold, T. and Vergunst, J.L. (2008). *Ways of Walking: Ethnography and Practice on Foot*. Aldershot: Ashgate.

Irvine, R. (2020). *An Anthropology of Deep Time*. Cambridge: Cambridge University Press.

Jamal, T. and Camargo, B. (2014). Sustainable tourism, justice and an ethic of care: Toward the just destination. *Journal of Sustainable Tourism* 22(1): 11–30.

Jamal, T. and Higham, J. (2021). Justice and ethics: Towards a new platform for tourism and sustainability. *Journal of Sustainable Tourism* 29(2–3): 143–157.

Jensen, A. (2009). From Thailand to Svalbard: Migration on the margins. *NIAS Nytt* 1: 13.

Jensen, A. and Moxnes, K. (2008). Livet i Longyearbyen: Åpne landskap – lukkede rom. In: Jensen, A. and Moxnes, K. (eds) *Livet i Longyearbyen: Åpne landskap – lukkede rom*, pp. 7–11. Trondheim: Tapir akademisk forlag.

Johannesen, S.-O. (2019). For mangfold og fellesskap. *Svalbardposten*, 10 September. At: https://svalbardposten.no/leserinnlegg/for-mangfold-og-fellesskap/19.11398 (accessed 5 June 2021).

Johnston, M., Viken, A. and Dawson, J. (2012). Firsts and lasts in Arctic tourism: Last chance tourism and the dialectic of change. In: Lemelin, H., Dawson, J. and Stewart, E.J. (eds) *Last Chance Tourism* (pp. 10–24). London: Routledge.

Joks, S., Østmo, L. and Law, J. (2020). Verbing meahcci: Living Sámi lands. *Sociological Review* 68(2): 305–321.

Jørgensen, A.-M. (2019). Communitification and emotional capital: Producing, shaping and re-shaping communities before and after mining in Norrbotten and Disko Bay. *Polar Record* 56: E7.

Junka-Aikio, L. and Cortes-Severino, C. (2017). Cultural studies of extraction. *Cultural Studies* 31(2–3): 175–184.

Kaltenborn, B.P. (2000). Arctic–Alpine environments and tourism: Can sustainability be planned? Lessons learned on Svalbard. *Mountain Research and Development* 20(1): 28–31.

Kaltenborn, B.P. and Emmelin, L. (1993). Tourism in the high North: Management challenges and recreation opportunity spectrum planning in Svalbard, Norway. *Environmental Management* 17(1): 41–50.

Kaltenborn, B.P., Østreng, W. and Hovelsrud, G.K. (2020). Change will be the constant – future environmental policy and governance challenges in Svalbard. *Polar Geography* 43(1): 25–45.

Kalvig, S. (2021). Verdens øyne er rettet mot Svalbard. *E24*, 20 April. At: https://e24.no/det-groenne-skiftet/i/x3083n/verdens-oeyne-er-rettet-mot-svalbard (accessed 1 July 2022).

Karlsen, M.A. (2018). The limits of egalitarianism: Irregular migration and the Norwegian welfare state. In: Bendixsen, S.K., Bringslid, M.B. and Vike, H. (eds) *Egalitarianism in Scandinavia: Historical and Contemporary Perspectives*, pp. 223–243. Cham: Palgrave Macmillan.

Karlsen, M.A. (2021). *Migration Control and Access to Welfare: The Precarious Inclusion of Irregular Migrants in Norway*. London: Routledge.

Klein, N. (2014). *This Changes Everything: Capitalism Vs. the Climate*. New York: Simon & Schuster.

Knox, H. (2015). Thinking like a climate. *Distinktion* 16(1): 91–109.

Kotašková, E. (2022). From mining tool to tourist attraction: Cultural heritage as a materialized form of transformation in Svalbard society. *Polar Record* 58: E19.

Kruse, F. (2013). *Frozen Assets: British Mining, Exploration and Geopolitics on Spitsbergen, 1904–53*. Groningen: Barkhuis.

La Cour, E. (2022). *Geo-aesthetical Discontent: Svalbard, the Guide and Post-Future Essayism*. PhD thesis, Göteborgs Universitet.

Laastad, S.G. (2021). The Janus face of local extractivism. *Extractive Industries and Society* 8(2), https://doi.org/10.1016/j.exis.2021.100903.

Landau, L.B. and Bakewell, O. (2018). Introduction: Forging a study of mobility, integration and belonging in Africa. In: Landau, L.B. and Bakewell, O. (eds) *Forging African Communities: Mobility, Integration and Belonging*, pp. 1–24. London: Palgrave Macmillan.

Latour, B. (2004). Why has critique run out of steam? From matters of fact to matters of concern. *Critical Inquiry* 30(2): 225–248.

Latour. B. (2014). Anthropology at the time of the Anthropocene – A personal view of what is to be studied. Distinguished lecture, American Association of Anthropologists, pp. 139-AAA 1–139-AAA 16. Available at: www.bruno-latour.fr/sites/default/files/139-AAA-Washington.pdf (accessed 9 Feburary 2023).

Lindberg, F. (2020). Uenighet om bærekraftig turisme. *Praktisk økonomi og finans* 36(2): 91–104.

Longyearbyen Lokalstyre (2013). *Lokalsamfunnsplan 2013–2023*. At: www.lokalstyre.no/lokalsamfunnsplan-2013-2023.504407.no.html (accessed 4 December 2022).

Longyearbyen Lokalstyre (2022). *Lokalsamfunnsplan 2023–2033*. At: https://img6.custompublish.com/getfile.php/5016489.2046.m7jzkspmlajbw7/2022.4.6+Lokalsamfunnsplan+til+sluttbehandling.pdf?return=www.lokalstyre.no (accessed 4 December 2022).

Lyons, N. (2010). The wisdom of elders: Inuvialuit social memories of continuity and change in the twentieth century. *Arctic Anthropology* 47(1): 22–38.

Malmo, V.K. (2021). Frykter for voldsutsatte kvinner på Svalbard: – Risikerer at det blir et fristed for voldsmenn. *NRK.no*, 19 April. At: www.nrk.no/tromsogfinnmark/frykter-for-voldsutsatte-kvinner-pa-svalbard_-_-risikerer-at-det-blir-et-fristed-for-voldsmenn-1.15451605 (accessed 5 June 2021).

Manzo, L.C., Williams, D.R., Raymond, Ch.M., Di Masso, A., Von Wirth, T. and Devine-Wright, P. (2021). Navigating the spaciousness of uncertainties posed by global challenges: A sense of place perspective. In: Raymond, Ch.M., Manzo, L.C., Williams, D.R., Di Masso, A., and Von Wirth, T. (eds) *Changing Senses of Place: Navigating Global Challenges*, pp. 331–347. Cambridge: Cambridge University Press.

Masco, J. (2015). Crisis in crisis. *Current Anthropology* 58(S15): S65–S76.

Massey, D. (1994). *Space, Place and Gender*. Cambridge: Polity Press.

Massey, D. (2004). Geographies of responsibility. *Geografiska Annaler: Series B, Human Geography* 86(1): 5–18.

McCright, A.M. and Dunlap, R.E. (2011). Cool dudes: The denial of climate change among conservative white males in the United States. *Global Environmental Change* 21(4): 1163–1172.

Meyer, A. (2022). Physical and feasible: Climate change adaptation in Longyearbyen, Svalbard. *Polar Record* 58: E29.

Midgley, S.J. (2012). *Co-producing Ores, Science and States: High Arctic Mining at Svalbard (Norway) and Nanisivik (Canada)*. Unpublished Master's thesis. St. John's: Memorial University of Newfoundland.

Milbourne, P. and Kitchen, L. (2014). Rural mobilities: Connecting movement and fixity in rural places. *Journal of Rural Studies* 34: 326–336.

Miller, E.C. (2021). *Extraction Ecologies and the Literature of the Long Exhaustion*. Princeton, NJ: Princeton University Press.

Milton, K. (1997). Ecologies: Anthropology, culture and the environment. *International Social Science Journal* 49(154): 477–495.

Moe, A. and Jensen, Ø. (2020). Svalbard og havområdene – nye utenrikspolitiske utfordringer for Norge? *Internasjonal Politikk* 78(4): 511–522.

Moen, E. (2021). *Hva kan vi lære av snøskredulykkene i Longyearbyen? En studie om innbyggerne og samfunnsutviklingen i Longyearbyen*. Unpublished student report. Longyearbyen: Longyearbyen skole.

Moore, A. (2015). Anthropocene anthropology: Reconceptualizing contemporary global change. *Journal of the Royal Anthropological Institute* 22: 27–46.

Moore, J. (2017). The Capitalocene, Part I: On the nature and origins of our ecological crisis. *Journal of Peasant Studies* 44(3): 594–630.

Moore, J. (2018). The Capitalocene Part II: Accumulation by appropriation and the centrality of unpaid work/energy. *Journal of Peasant Studies* 45(2): 237–279.

Moxnes, K. (2008). Familieliv i familiesamfunnet. In: Jensen, A. and Moxnes, K. (eds) *Livet i Longyearbyen: Åpne landskap – lukkede rom*, pp. 73–91. Trondheim: Tapir akademisk forlag.

Nhn.no (2020). Norsk helsenett i gang på Svalbard, 3 September. At: www.nhn.no/norsk-helsenett-i-gang-paa-svalbard/ (accessed 1 July 2022).

Nixon, R. (2011). *Slow Violence and the Environmentalism of the Poor*. Cambridge, MA: Harvard University Press.

Norgaard, K.M. (2011). *Living in Denial: Climate Change, Emotions, and Everyday Life*. Cambridge, MA: MIT Press.

Norwegian Ministry of Commerce (1991). *St. meld. nr. 50 (1990–1991): Næringstiltak for Svalbard*. At: www.stortinget.no/no/Saker-og-publikasjoner/ Stortingsforhandlinger/Lesevisning/?p=1990-91&paid=3&wid=d&psid= DIVL1575 (accessed 27 February 2023).

Norwegian Ministry of Defence (2021). *Høringsbrev – endring av valgbarhetsregler for valg til Longyearbyen lokalstyre m.m.* At: www.regjeringen.no/ contentassets/45eb0349d1b54c25ae1d4c634dea131d/forsvarsdepartementet. pdf?uid=Forsvarsdepartementet (accessed 3 December 2022).

Norwegian Ministry of Education and Research (2018). *Strategy for Research and Higher Education in Svalbard*. At: www.regjeringen.no/contentassets/ 3b322b7aec8942cf8a8bcd09e498547f/strategy-for-research-and-higher-education-in-svalbard.pdf (accessed 3 December 2022).

Norwegian Ministry of Justice (1986). *St. meld. nr. 40 (1985–1986): Svalbard*. At: www.stortinget.no/no/Saker-og-publikasjoner/Stortingsforhandlinger/ Lesevisning/?p=1985-86&paid=3&wid=c&psid=DIVL1260 (accessed 5 December 2022).

Norwegian Ministry of Justice and Public Security (1974–1975). *St. meld. nr. 39 (1974–1975): Vedrørende Svalbard*. At: www.stortinget.no/no/Saker-og-publikasjoner/Stortingsforhandlinger/Lesevisning/?p=1974-75&paid=3&wid= c&psid=DIVL814&s=True (accessed 3 December 2022).

Norwegian Ministry of Justice and Public Security (2015–2016). *Meld. St. 32 (2015–2016): Report to the Storting (White Paper)*. At: www.regjeringen.no/en/ dokumenter/meld.-st.-32-20152016/id2499962/ (accessed 30 March 2021).

Norwegian Ministry of Justice and Public Security (2021). *Endring av regler om stemmerett og valgbarhet ved valg til Longyearbyen Lokalstyre m.m.* At: www.regjeringen.no/contentassets/c5724aaf8cb545a39747629976cebf2e/ horingsnotat-endring-av-regler-om-stemmerett-og-valgbarhet-ved-valg-til-longyearbyen-lokalstyre-m.m..pdf (accessed 5 December 2022).

Nuttall, M. (2019). Sea ice, climate and resources: The changing nature of hunting along Greenland's northwest coast. In: Stensrud, A. and Eriksen, T.H. (eds) *Climate, Capitalism and Communities: An Anthropology of Environmental Overheating*, pp. 57–75. London: Pluto Press.

Nyseth, T. and Viken, A. (2016). Communities of practice in the management of an Arctic environment: Monitoring knowledge as complementary to scientific knowledge and the precautionary principle? *Polar Record* 52(1): 66–75.

Ødegaard, C.V. (2021). Sosiale drama på Svalbard: Tilbakeføring til natur og fortellinger om en ny tid. *Naturen* 145(2–3): 138–147.

Ødegaard, C.V. (2022). Returning to nature: Post-carbon utopias in Svalbard, Norway. *Social Analysis* 66(2): 1–22.

Ólafsdóttir, R. (2021). The role of public participation for determining sustainability indicators for Arctic tourism. *Sustainability* 13(1): 295.

Orlove, B., Wiegandt, E. and Luckman, B. (eds) (2008). *Darkening Peaks: Glacier Retreat, Science and Society*, pp. 229–240. Berkeley: University of California Press.

Øystå, F. (2022). Jørn har nyrekreft: – Giften er en bidragsyter. *Svalbardposten*, 27 June. At: www.svalbardposten.no/avinor-forurensing-miljodirektoratet/jorn-har-nyrekreft-giften-er-en-bidragsyter/486780 (accessed 1 July 2022).

Paglia, E. (2020). A higher level of civilisation? The transformation of Ny-Ålesund from Arctic coalmining settlement in Svalbard to global environmental knowledge center at 79° North. *Polar Record* 56: E15.

Pálsson, G. (1996). Human–environmental relations: Orientalism, paternalism and communalism. In: Descola, P. and Pálsson, G. (eds) *Nature and Society: Anthropological Perspectives*, pp. 65–81. London: Routledge.

Pasgaard, M., Fold, N., Meilby, H. and Kalvig, P. (2021). Reviewing tourism and natural resource research in the Arctic: Towards a local understanding of sustainable tourism in the case of South Greenland. *Geografisk Tidsskrift* 121(1): 15–29.

Pedersen, T. (2016). Gruvedrift og sikkerhetspolitikk. *Ottar* 310(2): 3–9.

Pedersen, T. (2017). The politics of presence: The Longyearbyen dilemma. *Arctic Review on Law and Politics* 8: 95–108.

Pedersen, T. (2021). The politics of research presence in Svalbard. *Polar Journal* 11(2): 413–426.

Pijpers, R. and Eriksen, T.H. (2018). *Mining Encounters*. London: Pluto Press.

Povinelli, E.A. (2016). *Geontologies: A Requiem to Late Liberalism*. New York: Duke University Press.

Quinn, T., Lorenzoni, I. and Adger, N. (2015). Place attachment, identity, and adaptation. In: O'Brien, K. and Selboe, E. (eds) *The Adaptive Challenge of Climate Change*, pp. 160–170. Cambridge: Cambridge University Press.

Rapp, O.M. (2019). Must adapt to a new reality. *Svalbardposten Special Edition: Top of the World. Living in the Arctic*, pp. 18–20.

Rastegar, R., Higgins-Desbiolles, F. and Ruhanen, L. (2021). COVID-19 and a justice framework to guide tourism recovery. *Annals of Tourism Research* 91: 103161.

Raymond, Ch.M., Manzo, L.C., Williams, D.R., Di Masso, A. and Von Wirth, T. (eds) (2021). *Changing Senses of Place: Navigating Global Challenges*. Cambridge: Cambridge University Press.

Ren, C., Jóhannesson, G., Kramvig, B., Pashkevich, A. and Höckert, E. (2021). 20 years of research on Arctic and Indigenous cultures in Nordic tourism: A review and future research agenda. *Scandinavian Journal of Hospitality and Tourism* 21(1): 111–121.

Revelin, F. (2013). Ecotourism and extraction in Saami lands: Contradictions and continuities. In: B. Büscher and V. Davidov (eds) *The Ecotourism/Extraction Nexus: Rural Realities and Political Economies of (Un)Comfortable Bedfellows*, pp. 193–214. London: Routledge.

Ritter, C. (2010 [1938]). *A Woman in the Polar Night*. Fairbanks, AK: University of Alaska Press.

Roberts, P. and Paglia, E. (2016). Science as national belonging: The construction of Svalbard as a Norwegian space. *Social Studies of Science* 46(6): 894–911.

Røsvik, H.K. (2019). Pliktoppfyllende ryddegutter. *Svalbardposten*, 7 March. At: www.svalbardposten.no/gruve-leder-opprydning/pliktoppfyllende-ryddegutter/177739 (accessed 1 July 2022).

Rudiak-Gould, P. (2011). Climate change and anthropology: The importance of reception studies. *Anthropology Today* 27(2): 9–12.

Rudiak-Gould, P. (2013). 'We have seen it with our own eyes': Why we disagree about climate change visibility. *Weather, Climate, and Society* 5: 120–132.

Rudiak-Gould, P. (2014). The influence of science communication on indigenous climate change perception: Theoretical and practical implications. *Human Ecology* 42: 75–86.

Ruiz-Frau, A., Kaiser, M.J., Edwards-Jones, G., Klein, C.J., Segan, D. and Possingham, H.P. (2015). Balancing extractive and non-extractive uses in marine conservation plans. *Marine Policy* 52: 11–18.

Ryan, A. (2022). *Polare kvinner. Norsk polarhistorie i kjønnsperspektiv.* Stamsund: Orkana Akademisk.

Saarinen, J. (2006). Traditions of sustainability in tourism studies. *Annals of Tourism Research* 33(4): 1121–1140.

Saarinen, J. (2014). Nordic perspectives on tourism and climate change issues. *Scandinavian Journal of Hospitality and Tourism* 14(1): 1–5.

Salazar, J.F., Pink, S., Irving, A. and Sjöberg, J. (eds) (2017). *Anthropologies and Futures: Researching Emerging and Uncertain Worlds.* London: Bloomsbury Academic.

Salazar, N. (2020). Labour migration and tourism mobilities: Time to bring sustainability into the debate. *Tourism Geographies* 24(1): 141–151.

Salem, T., Meyer, A. and Vindal Ødegaard, C. (in prep.) Fantasies and dilemmas of the empty wilderness.

Savage Minds (2015). *Anthropologies #21*: Is there hope for an Anthropocene anthropology?, 5 September. At: https://savageminds.org/2015/09/05/anthropologies-21-is-there-hope-for-an-anthropocene-anthropology/ (accessed 5 December 2022).

Saville, S. (2019a). The northern-most overtourism? At: https://samsaville.files.wordpress.com/2019/12/51fcd-the-northern-most-overtourism-for-website.pdf (accessed 5 December 2022).

Saville, S. (2019b). Tourists and researcher identities: Critical considerations of collisions, collaborations and confluences in Svalbard. *Journal of Sustainable Tourism* 27(4): 573–589.

Saville, S. (2022). Valuing time: Tourism transitions in Svalbard. *Polar Record* 58: E11.

Scott, J. (1998). *Seeing Like a State: How Certain Schemes to Improve the Human Condition Have Failed.* New Haven, CT: Yale University Press.

Sevestre, H., Douglas I.B., Luckman, A., Nuth, C., Kohler, J., Lindbäck, K. et al. (2018). Tidewater glacier surges initiated at the terminus. *Journal of Geophysical Research: Earth Surface* 123: 1035–1051.

Sillitoe, P. (2021). *The Anthroposcene of Weather and Climate.* New York: Berghahn Books.

Simonetti, C. (2019). Weathering climate: Telescoping change. *Journal of the Royal Anthropological Institute* 25(2): 241–264.

Simonetti, C. and Ingold, T. (2018). Ice and concrete: Solid fluids of environmental change. *Journal of Contemporary Archaeology* 5(1): 19–31.

Singleton, B., Gillette, M., Burman, A. and Green, C. (2021). Toward productive complicity: Applying 'traditional ecological knowledge' in environmental science. *Anthropocene Review*, online first: 1–22.

Sisneros-Kidd, A.M., Monz, C., Hausner, V., Schmidt, J. and Clark, D. (2019). Nature-based tourism, resource dependence, and resilience of Arctic communities: Framing complex issues in a changing environment. *Journal of Sustainable Tourism* 27(8): 1259–1276.

Skagestad, J. (2021). Svalbard – Canary in the coal mine goes green. *TEDxSkift*. At: www.youtube.com/watch?v=bb-xwjKKirE (accessed 1 July 2022).

Skaptadóttir, U. (2010). Integration and transnational practices of Filipinos in Iceland. *e-Migrinter* 5: 35–46.

Skaptadóttir, U. (2019). Transnational practices and migrant capital: The case of Filipino women in Iceland. *Social Inclusion* 7(4): 211–220.

Skaptadóttir, U. and Innes, P. (2017). Immigrant experiences of learning Icelandic and connecting with the speaking community. *Nordic Journal of Migration Research* 7(1): 20–27.

Skjæraasen, M., Helledal, E.J. and Klausen, D.H. (2021). Måtte reparere frøhvelvet for 193 millioner. Skyldte på klimaendringer. *NRK.no*, 22 March. At: www.nrk.no/klima/matte-reparere-frohvelvet-for-193-millioner.-skyldte-pa-klimaendringer.-1.15412828 (accessed 1 July 2022).

Snow, C. (2012 [1959]). *The Two Cultures*. Cambridge: Cambridge University Press.

Søholt, S., Stenbacka, S. and Nørgaard, H. (2018). Conditioned receptiveness: Nordic rural elite perceptions of immigrant contributions to local resilience. *Journal of Rural Studies* 64: 220–229.

Sokolíčková, Z. and Soukupová, E. (2021). Czechs and Slovaks in Svalbard: Entangled modes of mobility, place and identity. *Urban People* 23(2): 167–196.

Sokolíčková, Z., Meyer, A. and Vlakhov, A. (2022). Changing Svalbard: Tracing interrelated socioeconomic and environmental change in remote Arctic settlements. *Polar Record* 58, E23.

Sokolíčková, Z., Ramirez Hincapié, E., Zhang, J., Lennert, A.E., Löf, A. and van der Wal, R. (2023). Waters that matter: How human–water relations are changing in High Arctic Svalbard. *Anthropological Notebooks* 28(3): 74–109.

Solberg, M.F. (2019). Gøy å se at det er ferdig allerede nå. *Svalbardposten*, 13 July. At: www.svalbardposten.no/boligbygging-boliger-boligmarkedet/goy-a-se-at-det-er-ferdig-allerede-na/441646 (accessed 1 July 2022).

Solberg, M.F. (2020). Dette skulle styrkes – fortsatt en lang vei å gå. *Svalbardposten*, 20 February. At: www.svalbardposten.no/lokalstyret-longyearbyen-lokalstyre-norsk/dette-skulle-styrkes-fortsatt-en-lang-vei-a-ga/189572 (accessed 1 July 2022).

Sörlin, S. (2021). Wisdom of affect? Emotion, environment, and the future of resource extraction. *Polar Record* 57: E27.

Sörlin, S. (2023). The new extractivist paradigm. In: Sörlin, S. (ed.) *Resource Extraction and Arctic Communities: The New Extractivist Paradigm*, pp. 3–32. Cambridge: Cambridge Ubiversity Press.

Statistics Norway. (2021). *Population of Svalbard*, 4 May. At: www.ssb.no/en/befolkning/folketall/statistikk/befolkningen-pa-svalbard (accessed 4 June 2021).

Stensrud, A. (2016). 'We are all strangers here': Transforming land and making identity in a desert boomtown. In: Eriksen, T.H. and Schober, E. (eds) *Identity Destabilised: Living in an Overheated World*, pp. 59–76. London: Pluto Press.

Stensrud, A. and Eriksen, T.H. (2019). *Climate, Capitalism and Communities: An Anthropology of Environmental Overheating*. London: Pluto Press.

Stoddart, M.C., Mattoni, A. and McLevey, J. (2020). Introduction: Contact points between offshore oil and nature-based tourism. In: Stoddart, M.C., Mattoni, A. and McLevey, J. (eds) *Industrial Development and Eco-tourisms: Can Oil Extraction and Nature Conservation Co-exist?*, pp. 1–26. Cham: Springer International Publishing.

Strømmen, K. (2016). *Longyearbyen skoles historie*. Longyearbyen: Longyearbyen Lokalstyre.

Thisted, K., Sejersen, F. and Lien, M.E. (2021). Arctic uchronotopias: Resource extraction, community making and the negotiation of Arctic futures. *Polar Record* 57: E28.

Tiller, R.G., Ross, A.D. and Nyman, E. (2022). Social capital and institutional complexity in Svalbard: The case of avalanche disaster management. *Disaster Prevention and Management* 31(4): 425–439.

Tönnies, F. (2001 [1887]). *Community and Civil Society*. Cambridge: Cambridge University Press.

Traa, K., Sande, T. and Soland, J. (2022). Luftslott om det grønne skiftet i Longyearbyen. *Svalbardposten*, 2 May. At: www.svalbardposten.no/energi-energiforsyning-leserinnlegg/luftslott-om-det-gronne-skiftet-i-longyearbyen/213069 (accessed 1 July 2022).

Tsing, A. (2015a). More-than-human sociality: A call for critical description. In: Hastrup, K. (ed.) *Anthropology and Nature*, pp. 27-42. New York: Routledge.

Tsing, A. (2015b). *The Mushroom at the End of the World: On the Possibility of Life in Capitalist Ruins*. Princeton, NJ: Princeton University Press.

Tsing, A. (2019). The political economy of the Great Acceleration, or, How I learned to stop worrying and love the bomb. In: Stensrud, A. and Eriksen, T. (eds) *Climate, Capitalism and Communities: An Anthropology of Environmental Overheating*, pp. 22–40. London: Pluto Press.

Uusiautti, S. and Yeasmin, N. (2019). *Human Migration in the Arctic: The Past, Present, and Future*. Singapore: Palgrave Macmillan.

Vertovec, S. (2007). Super-diversity and its implications. *Ethnic and Racial Studies* 30(6): 1024–1054.

Vertovec, S. (2023). *Superdiversity: Migration and Social Complexity*. New York: Routledge.

Viken, A. (1998). Miljødiskurs på Svalbard som akademisk hegemoni. *Sosiologi i dag* 28(2): 83–109.

Viken, A. (2006). Svalbard. In: Baldachino, G. (ed.) *Extreme Tourism: Lessons from the World's Cold Water Island*, pp. 128–142. London: Routledge.

Viken, A. (2011). Tourism, research and governance on Svalbard: A symbiotic relationship. *Polar Record* 47: 335–347.

References 195

Viken, A. (2020a). Turismens vesen og uvesen. In: Viken, A., Benonisen, R., Ekeland, C., Førde, A., Nilsen, R., Nyseth, T. et al. (eds) *Turismens paradokser: Turisme som utvikling og innvikling*, pp. 11–25. Stamsund: Orkana Akademisk.
Viken, A. (2020b). Turisme, kunnskapshegemonier og sjølregulering på Svalbard. In: Viken, A., Benonisen, R., Ekeland, C., Førde, A., Nilsen, R., Nyseth, T. et al. (eds) *Turismens paradokser: Turisme som utvikling og innvikling*, pp. 301–320. Stamsund: Orkana Akademisk.
Viken, A. and Jørgensen, F. (1998). Tourism on Svalbard. *Polar Record* 34(189): 123–128.
Viken, A., Johnston, M., Nyseth, T. and Dawson, J. (2014). Responsible Arctic tourism: Myth or reality? A case study of Svalbard and Nunavut. In: Viken, A. and Granås, B. (eds) *Destination Development in Tourism: Turns and Tactics*, pp. 245–261. London: Routledge.
Visit Svalbard (2022a). Revidert Masterplan for Svalbard mot 2030. At: www.visitsvalbard.com/dbimgs/RevidertmasterplanSvalbard2mai2022.pdf (accessed 1 July 2022).
Visit Svalbard. (2022b). Increased efforts on sustainable tourism. At: https://en.visitsvalbard.com/inspiration/various/increased-efforts-sustainable-tourism (accessed 1 July 2022).
Wiersen, S.Ø. (2020). Vi skal ha god kontroll på hvem som bor her. *Svalbardposten*, 25 February. At: www.svalbardposten.no/vi-skal-ha-god-kontroll-pa-hvem-som-bor-her/189836 (accessed 5 December 2022).
Winther, J.-G. (2015). The Arctic: A global climate 'canary in a coal mine', 27 May. At: www.openaccessgovernment.org/arctic-global-climate-canary-coal-mine/17661/ (accessed 1 July 2022).
WWF Norway (2010). Høringssvar 'Ny kullgruve i Lunckefjell – høring av konsekvensutredning og søknad om tillatelse etter Svalbardmiljøloven'. At: https://web.archive.org/web/20160306023552/http://assets.wwf.no/downloads/wwf_horingssvar___ny_kullgruve_i_lunckefjell.pdf (accessed 1 July 2022).
Ylvisåker, L.N. (2016). Føler de er blitt utnyttet. *Svalbardposten*, 23 June At: https://svalbardposten.no/nyheter/foler-de-er-blitt-utnyttet/19.7585 (accessed 5 June 2021).
Ylvisåker, L.N. (2020). Uvanleg sommar, 23 September 2020. At: www.ylvisaaker.com/new-page-4 (accessed 1 July 2022).
Ylvisåker, L.N. (2022). *My World is Melting*. Longyearbyen: Ylvisåker AS.

Index

Thanks to our Patreon subscriber:

Ciaran Kane

Who has shown generosity and
comradeship in support of our publishing.

Check out the other perks you get by subscribing
to our Patreon – visit patreon.com/plutopress.

Subscriptions start from £3 a month.